DATE DUE

ILL: 9306955	SAZ	
JUL 2 5		
Aug 7		
AUG 20		
Sept 2		

The Conscious Universe

Menas Kafatos   Robert Nadeau

The Conscious Universe

Part and Whole in Modern Physical Theory

With 31 Illustrations

Springer-Verlag
New York Berlin Heidelberg London Paris
Tokyo Hong Kong Barcelona Budapest

Menas Kafatos
Department of Physics
George Mason University
Fairfax, VA 22030-4444
USA

Robert Nadeau
Department of English
George Mason University
Fairfax, VA 22030-4444
USA

Original artwork for the cover and figure illustrations created by M. Kafatos, generated on the Imaginator D-80 design station, and imaged on the D148-SR Crosfield/Dicomed Film Recorder at Blair Inc. Art director for the project was Jackie Hancock and the computer graphics artist was Chris Engnoth of Blair Inc., Alexandria, VA.

Kafatos, Menas
    The conscious universe : part and whole in modern physical theory
    / Menas Kafatos, Robert Nadeau.
        p.   cm.
    Includes bibliographical references.
    ISBN 0-387-97262-5
    1. Science—Philosophy. 2. Physics—Philosophy. 3. Quantum
theory. I. Nadeau, Robert, 1944- . II. Title.
    Q175.K13   1990
    501—dc20   90-31417

Printed on acid-free paper.

Text prepared on Ventura Publisher using author-supplied WordPerfect disk.
Printed and bound by R.R. Donnelley & Sons, Harrisonburg, VA.
Printed in the United States of America.

9 8 7 6 5

ISBN 0-387-97262-5 Springer-Verlag New York Berlin Heidelberg
ISBN 3-540-97262-5 Springer-Verlag Berlin Heidelberg New York

To Thalia and Sonja

Contents

Introduction

This discussion resulted from a dialogue which began some seven years ago between a physicist who specializes in astrophysics, general relativity, and the foundations of quantum theory, and a student of cultural history who had done post-doctoral work in the history and philosophy of science. Both of us at that time were awaiting the results of some experiments being conducted under the direction of the physicist Alain Aspect at the University of Paris-South.[1] The experiments were the last in a series designed to test some predictions based on a mathematical theorem published in 1964 by John Bell.[2] There was no expectation that the results of these experiments would provide the basis for developing new technologies. The questions which the experiments were designed to answer concerned the relationship between physical reality and physical theory in the branch of physics known as quantum mechanics. Like most questions raised by physicists which lead to startling new insights, they were disarmingly simple and direct.

Is quantum physics, asked Bell, a self-consistent theory whose predictions would hold in a new class of experiments, or would the results reveal that the apparent challenges of quantum physics to the understanding in classical physics of the relationship between physical theory and physical reality were merely illusory? Answering this question in actual experiments could also, suggested Bell, lead to another, quite dramatic, result. The character of physical reality as predicted by quantum physics was non-local, meaning that correlations between results would hold in spite of the fact that the regions in space in which they would be observed would be too distant from one another to allow signals traveling at the speed of light to travel between the two points in the time allowed. Or, the character of physical reality would violate the predictions of quantum physics and would, therefore, be local, meaning that these correlations would not be observed in accordance with predictions made by classical physics. When we finally saw the results of the Aspect experiments, published in 1982, the answers seemed quite clear—quantum physics was a self-consistent theory, and the character of physical reality as disclosed by quantum physics is non-local.

Modern physics has, of course, consistently provided us with an understanding of the character of physical reality at odds with our everyday sense of this reality. But no previous discovery has posed more challenges to our sense of everyday reality than the discovery that non-locality is a fact of nature. Non-locality is a shocking discovery because it appears to subvert the bias that the world is composed most fundamentally of individual objects and their non-relational properties. The apparent inability of objects to evince non-relational properties when they are separate from one another seems self-evident in every action we undertake. If this were not so, we could not act or react to anything in the macro world with the sense that there is either discreteness or causality.

1

Although each of us had followed the experiments testing Bell's theorem for different reasons, we were both convinced that these results represented another signal event in a revolution in thought that was extending itself well beyond the narrow concerns of quantum physicists or philosophers and historians of science. After the results were published, some physicists seemed initially preoccupied with the suggestion that the correlations between measurements of the paired photons, or quanta of light, in these experiments were occurring at a superluminar speed, or at a speed faster than light. Our view was that Bell's theorem and the Aspect experiments testing the theorem did not violate relativity theory, and that the solution to the enigma of non-locality had nothing to do with faster-than-light signals.

More important both of us also intuited, not knowing where this intuition would lead, something that was not terribly obvious at the time. It appeared as if these results had provided final confirmation that the classical view of the relationship between physical theory and physical reality, which quantum physics had been challenging for some time, was no longer supportable. If this was the case, the classical conception of the relationship between part and whole in physical theories was also, we felt, quite wrong. As the dialogue continued, we discovered that the dual perspective provided by our diverse academic backgrounds was absolutely essential in making sense out of the new situation. We also found ourselves appealing increasingly to the little-read commentary of the Danish physicist Niels Bohr on the implications of the quantum mechanical description of nature, and used it as our guide in moving into previously unexplored territory. Sensing that the conclusions we were drawing could be quite important not only for the future of physical theories, but also for an improved understanding of the relationship between the truths of science and those in other aspects of human experience, the challenge became to put the discussion in a form that might serve both ends.

During the period in which we were writing and rewriting the present book, we attended conferences which dealt with related issues, and entered into dialogue with other physicists, as well as historians and philosophers of science, who shared our concerns. Eventually, we found ourselves organizing an international conference on Bell's theorem and the related experiments which met in October of 1988, and which was well attended by major figures in all areas.[3] The contributions of these individuals proved invaluable to our own efforts.

The decision to write a book for the general reader, as opposed to specialists, was motivated by our conviction that the revolution in progress was simply too important to leave its future merely in the hands of specialists. A profound revision in our understanding of the character of scientific truth would obviously be a major development. Since the eighteenth century, science and its methods have played a large role in legitimating all forms of knowledge. If one believes that scientific knowledge has more authority in the description of the universe than other ways of knowing, assumptions about the ultimate foundations of truth are now in need, as we shall see, of some drastic revisions. More important, the new relationship between part and whole disclosed in modern physical theories, and grandly reinforced by Bell's theorem and the experiments testing that theorem, could carry implications that might help us more realistically comprehend our actual relation to one another in this dangerously divided world.

Physics for non-physicists books seem to fall rather neatly these days into two categories. In the first category, physicists, or those with more than a passing acquaintance with physics, attempt to translate into ordinary language some understanding of the character of physical reality disclosed in modern physics. In the second category of physics for non-physicists books, a good deal of emphasis is placed on the potential impact of modern physics on human values and institutions, and there is often the suggestion in such books that the implications of modern physics could or should occasion a revolution in metaphysics. Where metaphysics is a primary emphasis in books in this second category, the usual conclusion seems to be that the world view of modern physics is more consistent with eastern metaphysics, particularly Taoism, Hinduism, and Buddhism. Although some writers who have drawn parallels between modern physics and eastern metaphysics are well known and respected physicists, like Fritjof Capra or David Bohm, the majority of physicists tend not to be terribly impressed by these efforts. The obvious explanation for this reaction is that most physicists seem to be convinced that physics has nothing to do with metaphysics, and that any attempt to force a dialogue between the two can only result in dangerous and groundless speculation. This discussion does not fall into either category of physics for non-physicists books, or more accurately in our view, extends itself beyond both.

Although we will make the case that metaphysical assumptions and presuppositions have played a role in the history of science, and continue to play such a role in what we will term here the "hidden ontology of classical epistemology," we also adopt the position that metaphysics should, ideally at least, have nothing whatsoever to do with the actual practice of physics. And yet we will also make the case that the discovery that non-locality is a new fact of nature allows us to "infer," although certainly not to "prove," that the universe can be viewed as a conscious system. Since these views would normally be taken as incompatible and logically inconsistent, the manner in which we have sought to make them self-consistent obviously requires some explanation. Before we provide that explanation, it will first be necessary to articulate our views on the character of scientific truths, on the manner in which the truths of mathematical physics evolve, and on the relation of these truths to those in other aspects of human experience.

Science in our view, and in the view of virtually all physical scientists, is, first of all, a rational enterprise committed to obtaining knowledge about the actual character of physical reality. And the only way in our view to properly study the history and progress of science is to commit oneself to metaphysical and epistemological realism. In the view of Niels Bohr, which most physicists would probably accept, the universe is presumed real independently of human observers or any acts of observation. We term this view of the universe metaphysical realism. Epistemological realism, on the other hand, requires strict adherence to and regard for the rules and procedures for doing science. In the case of quantum physics, where we confront some very unique epistemological problems, a commitment to epistemological realism also requires, as we understand it, that we accept the proposition that any phenomenon can be presumed a "real" phenomenon only when it is "observed phenomenon," and that no physical theory can be presumed valid unless its predictions are subject to proof in repeatable scientific experiments under controlled conditions. In other words, the validity of any physical theory is dependent on acts of observation which disclose particular aspects of physical reality. It is this commitment to epistemological and metaphysical realism, as we hope to demonstrate, that allows us to examine the manner in which psychological and social phenomena have always conditioned by varying degrees the actual progress of science.

At the same time our definitions of epistemological and metaphysical realism differ markedly, for reasons we will explore in a moment, from those of members of the "Vienna circle," or of figures like Hans Hahn, Moritz Schlick, Philip Frank, and Rudolph Carnap. What these thinkers sought to demonstrate was that all of science could be unified by relying "exclusively" on the language of mathematics or, more accurately, that of symbolic logic. Such a procedure allows one, they argued, to conclude that all philosophical questions, including those relating to epistemology, can be completely eliminated from the practice of science. Although many historians of science have demonstrated that philosophical concerns and epistemological assumptions and problems do, in fact, loom very large in the history of scientific thought, the position taken by members of the Vienna school has been rather effectively undermined for other reasons as well. The two most obvious are that the epistemological problems confronted in quantum physics cannot be resolved with an appeal to symbolic logic alone, and the demonstration by Kurt Gödel in his famous "proof" that no mathematical system can "in itself" be complete. Gödel's theorem, which we will appeal to late in this discussion, demonstrates that no algorithm that demonstrates a mathematical proof can also prove its own validity. In order to provide such a proof, a larger and more embracing algorithm is required which, in turn, cannot prove its own validity, and so on.

It seems fair to say, however, that most educated people are very much in agreement with one assumption made by members of the Vienna circle that has always appeared rather self-evident to physical scientists. That assumption is that the description of physical reality provided by mathematical physics is far more capable of disclosing the actual character of this reality than descriptions framed in ordinary language. Yet there are many well-educated humanists and social scientists, including some philosophers of science, who have adopted a set of assumptions and attitudes about the character of scientific truths and the manner in which they originate and evolve which serve to either greatly diminish their authority or, in the extreme case, to render such truths as virtually irrelevant to the pursuit of knowledge. Although the arguments from the philosophy of science that have served in varying degrees to bolster these assumptions and attitudes, advanced by figures like Toulmin, Kuhn, Hanson, and Feyerabend, are far too detailed and complex to be reviewed in any detail here, we can at least provide some sense of why we find them less than persuasive. If the reader is a philosopher of science and wishes to read a detailed account of the inadequacies of these arguments from our perspective, Frederick Suppe in the second edition of *The Structure of Scientific Theories* provides such an account.[4]

All of these philosophers, as Suppe suggests, operate on the assumption that science is done within the context of a *Weltanschauung*, or comprehensive world-view, which is a product of culture and primarily resident in ordinary or linguistically-based language.[5] Although one would be foolish to discount this view entirely, as we clearly do not in our brief histories of mathematical physics, it can, if taken to extremes, lead to some rather untenable and even absurd conclusions about the progress of science and its epistemological authority vis-à-vis other knowledge fields.

Toulmin's primary argument is that science functions within a *Weltanschauung* to build up systems of ideas about nature which make some legitimate claim to "reality." We take such a claim as legitimate, he suggests, when the systems provide "explanatory techniques" which are consistent with the data and "pleasing to the mind."[6] Theories are, he asserts, designed to "predict" in the sense that they "provide explanations of recognized regularities."[7] Although the theories consist of laws and hypotheses, both exist, claims Toulmin, in a hierarchical structure in which "ideals of natural order" provide the whole orientation to the subject.[8] Since theories in this view are rules for drawing inferences, they can be neither true nor false—they simply work or do not work. *Weltanschauung* enters the picture here in the form of linguistically-based "ideals of natural order" which determine what questions are asked by the scientist, and, therefore, the assumptions that underlie his or her theorizing.[9]

What Toulmin implies, of course, is that cultural change can, in concert with changes in scientific knowledge, alter the "ideals of natural order," and thereby impact the assumptions that underlie the theorizing of the scientist as well as what he or she chooses to regard as scientific facts. Although there can be no argument that theories have a good deal to do with linguistically-based presumptions of scientists, particularly their "ideals of order," theories are not merely rules for drawing inferences about phenomena. As Suppe puts it, "there are certain charac-teristics typical of laws which are inapplicable to rules of inference. First, it is possible to give evidence in support of laws, but it is impossible to do so for rules: the closest you can come for giving evidence for rules is to give evidence that they are obeyed or in support of the claim that they should be accepted; but neither of these give evidence for the rules themselves in any way analogous to the way one can give evidence for laws."[10] If we do not recognize or understand this distinction between rules and laws, it becomes possible to presume that there is an extreme instrumentalism in the history of scientific thought in which the wayward motions of *Weltanschauung* play a rather primary role. Those who have read Toulmin in this way have occasionally drawn a conclusion far more extreme than that of Toulmin himself—since the course of science is driven primarily by the vagaries of cultural change, the notion that science is a context-driven and cumulative way of knowing featuring a high level of internal consistency must be false. This, as we will try to demonstrate in the course of this discussion, is a totally untenable position, even though extra-scientific or metaphysical assumptions have played and continue to play a role in the history of science.

Those who wish to dismiss the notion that science is essentially cumulative have more typically appealed to Kuhn's widely read *The Structure of Scientific Revolutions*. Here Kuhn argues that the evolution of *Weltanschauung* can be rather discontinuous in that there have been occasions when new ideals of natural order are rather quickly rejected in favor of other such ideals. Whereas Toulmin suggests, quite rightly we think, that scientific advances are cumulative, and cannot, therefore, be truly revolutionary, Kuhn in *The Structure of Scientific Revolutions* took the opposite view. His thesis, in brief, is that "scientific revolutions are ... those non-cumulative development episodes in which an older paradigm is replaced in whole or in part by an incompatible new one."[11] Kuhn defines paradigms as,

> accepted examples of actual scientific practice—examples which include law, theory, application, and instrumentation together—[which] provide models from which spring particular coherent traditions of scientific research.[12]

Periods of normal science in which an accepted paradigm is well established are punctuated by "paradigm shifts," or scientific revolutions, which can be viewed as analogous to what psychologists term a "gestalt switch."

Many humanists and social scientists have continued to appeal to *The Structure of Scientific Revolutions*, particularly given the fact that Kuhn is also a trained physicist, as a convincing demonstration that science is not concerned with uncovering the truths of nature, and that what passes as such truths are merely what the scientific community allows within the confines of regnant or competing paradigms. Some have even gone so far as to suggest that Kuhn has shown that science is an irrational enterprise governed and directed by waywardly evolving social institutions.[13] Although there is very little textual evidence to suggest that Kuhn views science as an irrational enterprise, and he has himself taken great pains to disallow any suggestion that science is irrational, there is little doubt that some massive misconceptions of the nature of scientific thought and progress can be traced to *The Structure of Scientific Revolutions*. Those who continue to reinforce these misconceptions with an appeal to this study should be advised that Kuhn himself, in response to some rather scathing criticisms of his thesis, has substantially revised his earlier position. He has given up on the notion that paradigms are anything like overarching world views, and that the history of science is driven by massive discontinuities.[14]

Hanson's *Weltanschauung* argument begins with the assumption that any scientific view is theory-laden, and thus physical reality as it is viewed by the scientist is seen through a conceptual pattern. One aspect of any theory is, he claims, the function of the meanings attached to terms within a linguistically-based context, and another is a function of the lawlike generalizations, hypotheses, and methodological approaches held in context.[15] Thus a theory, says Hanson, is a pattern of conceptual organization that explains phenomena by making them intelligible, and theories developed to explain unexplained phenomena are limited to or circumscribed by one's pattern of conceptual organization.

The most unique aspect of Hanson's argument, which is based on his understanding of the role played by linguistically-based elements in physical theory, concerns the logic of scientific discovery. The fundamental question he raises is whether two scientists who hold radically different theories about a particular phenomenon "see" the phenomenon in the same way. His answer is that they do not, and he offers some historical examples, like that of Kepler and Tycho Brahe, in support of this conclusion. The primary motive here is to dethrone the notion that there is an intersubjective language used by scientists that leads to an "objective" account of phenomena because there are linguistically-based elements in all theories which disallow the notion that a theory can be "neutral." This linguistically-based element conditions, claims Hanson, what is "seen" as the scientist "looks" for "a conceptual pattern in terms of which his data will fit intelligibly along better known data."[16] This allows one to conclude, he suggests, that theories are not discovered based on inductive generalizations from data but rather by "retrospectively inferring probable hypotheses from conceptual organized data."[17] What tends to convey the false impression that new theories derive exclusively from inductive generalizations is what Hanson calls "retroductive reasoning," a process in which the true logic of scientific discovery, which invokes theory-laden and linguistically-based perception, is disguised when scientists appeal to neutral observation language to describe their results.

In our view this position is reasonable, and we will offer examples from the history of science later in the discussion which tend to support it. Our essential disagreement with Hanson involves, however, what appears to be an overemphasis on the linguistic factor in "seeing" which has led some of his readers to the conclusion that linguistically-based theory-laden perception is more fundamental, primary, or causal in research in physics than is actually the case. If one places too much emphasis on the overemphasis, then it is possible to conclude that Hanson is advocating along with Kuhn the view that *Weltanschauung* drives the progress of science, and that the truths of science lack, therefore, the epistemological authority which neutral observation language has served to disguise.

The last of the *Weltanschauung* theorists we have chosen to discuss here is the most radical—Feyerabend. In contrast with positions taken by earlier philosophers of science like Popper, Feyerabend takes the view that there is no privileged method of scientific inquiry which leads to the successful acquisition of scientific knowledge. He advocates "methodological anarchy," and even characterizes himself as a Dadaist.[18] Science in his view has no special method of its own that makes it a privileged way of knowing, and should not be regarded as a strictly rational enterprise. Progress in science occurs, he asserts, when scientists think "counter-intuitively," and make radical departures from previously practiced norms of thought.

Although there are no physicists in our acquaintance who would even begin to completely embrace Feyerabend's views, the same cannot be said of many humanists and social scientists. Since Feyerabend suggests that no particular form of knowledge, including all extant methodologies, can claim a privileged status, some humanists and social scientists who apparently sense that their knowledge fields and associated methodologies are sometimes viewed as inferior to those of the hard sciences appear to have found Feyerabend's attempt to undermine the privileged status of scientific knowledge rather appealing. Those who derive comfort from Feyerabend in this manner should perhaps know that he has also advocated, based on the presumption that the privileged position of science within the educational curriculum be abandoned, that "special creationism" should be taught as an elective along with evolutionary biology. The only positive thing we can personally say about Feyerabend's theorizing is that rationality does take myriad forms, and that intuitive and seemingly irrational acts of the imagination have on occasion contributed to the overall rational progress of science.

It is fair to say that the *Weltanschauung* theorists have not been taken very seriously by the community of physicists, and also that this entire approach to the philosophy of science has been largely displaced by what Suppe terms "historical realism." This approach is characterized, according to Suppe, by "paying close attention to actual scientific practice, both historical and contemporary, all in the aim of developing a systematic philosophical understanding of the justification of knowledge claims."[19] If we were to identify ourselves with the views of any figures in the philosophy of science who practice historical realism, that figure would be Dudley Shapere.

Shapere's focus is on the reasoning patterns in actual science, and on the manner in which physics as a "privileged" form of coordinating experience with physical reality has obliged us to change our views of ourselves and the universe. We are also in full agreement with Shapere's view that the cumulative progress of science does impose constraints on what can be viewed as a legitimate scientific concept, problem, or hypothesis, and that these constraints become "tighter" as science progresses. This is particularly so, as we hope to demonstrate, when the results of theory present us with radically new and seemingly counterintuitive findings like those of the Aspect experiments. It is because, as Shapere notes, there is incessant "feedback" within the content of science that we are led to findings like that of non-locality as a fact of nature. This also means that the postulates of rationality, generalizability, and systematizability are rather consistently vindicated in the history of science.[20] Although Shapere does not, as we do not, dismiss the prospect that theory and observation can be and are conditioned by extra-scientific, linguistically-based factors, he argues, correctly in our view, that this does not finally compromise the objectivity of scientific knowledge. This means that although the psychological and sociological context of the scientist is an important aspect of the study of the history and evolution of scientific thought, the progress of science and its epistemological validity is not ultimately directed or governed by such considerations.

It is our commitment to "historical realism," as we hope to demonstrate in the chapters dealing with the history of physical theories, that allows us to conclude that much of the confusion about the character of physical reality that currently exists among physical scientists, particularly in regard to epistemological assumptions about the relationship between physical theory and physical reality, can be traced to some metaphysical assumptions and presuppositions in western thought.

Similarly, it is this same commitment in the examination of the conditions and results of the Aspect experiments which leads to the conclusion that non-locality "infers" the existence of an undivided wholeness on the primary level in physical reality without in any sense being able to "prove" its existence. If we do, in fact, live in a quantum mechanical universe, and we will devote some space in this discussion to demonstrating that this is the case, then the conditions and results of these experiments clearly suggest that there is a new and unanticipated limit on the ability of physical theory to disclose or describe all aspects of physical reality.

Although it will not be possible to fully appreciate why this is the case prior to our examination of the conditions and results of the Aspect experiments, it has to do with one of the assumptions regarding epistemological realism in quantum physics noted earlier—no phenomenon can be presumed to be a real phenomenon until it is an observed phenomenon. What is both interesting and amazing about non-locality is that it cannot be viewed "in principle" as an observed phenomenon. What we "observe" in the Aspect experiments are correlations between properties of quanta of light, or photons, emanating from a single source based on "measurements" made in space-like separated regions, or at two points that are so distant from one another that the correlations occur in apparent violation of the upper limit at which signals can travel in the universe—the speed of light. Yet what cannot be measured or observed "in principle" in this experimental situation is the reality that exists between the two points. What makes this fact even more dramatic in its implications, as we will attempt to explain in some detail later, is that all quanta can be assumed to have interacted at some point in the history of the cosmos in the manner that quanta interact at the source of origins in the Aspect experiments. What this means, in short, is that non-locality can be assumed to be a fundamental property of the entire universe. Thus the reality whose existence is inferred between the two points in the Aspect experiments should properly be viewed, as we understand it, as a reality that underlies and informs all physical events in the universe. And yet all that we can say about this reality is that it appears to be an indivisible whole whose existence is "inferred" where there is an interaction with an observer, or with instruments of observation, and that it also appears to exist outside of or beyond space-time.

When we consider that an indivisible whole by definition contains no separate parts, and concede that a phenomenon can be assumed to be real only when it is an "observed" phenomenon, then we are led here to some interesting conclusions. The indivisible whole whose existence is inferred in the Aspect experiments, and which also seems to be a property of the entire universe, cannot "in principle" be itself the subject of scientific investigation and lies, therefore, outside of the domain of scientific knowledge. The simple explanation as to why this is the case derives from adherence to epistemological realism—science can claim knowledge of physical reality only when the predictions of a physical theory are validated by experiment, and experiments necessarily involve the measurement of quantities describing actual properties in terms of space and time, or more properly in modern relativistic physics, in terms of the space-time continuum. Since the indivisible whole whose existence is inferred in the Aspect experiments cannot "in principle" be measured or quantified and also appears to exist outside of or beyond space-time, we confront here an "event horizon" of knowledge where science can say nothing about the actual character of this reality.

If we assume that this reality is also a property of the entire universe at all times and scales, then we can also infer that an undivided wholeness exists on the most primary and basic level in physical reality. Thus what we are actually dealing with in science per se are manifestations of this reality which are invoked or "actualized" in making acts of observation or measurement. And this leads to another, and even more extraordinary, conclusion—the sum of the parts cannot "in principle" be assumed to constitute this "indivisible" whole, and, therefore, physical theory cannot even "in principle" disclose or describe its actual character. Whether one chooses to regard this indivisible whole as having an ontological dimension is, of course, a matter of personal belief or conviction. And yet it seems clear that any ontological or metaphysical questions that we might choose to raise regarding this indivisible whole, or what we have chosen to call reality-in-itself, cannot be legislated over by the truths of science for the reason we have already noted—this reality cannot "in principle" be disclosed or described by scientific theory or experiment.

What we are suggesting here is that those of us who do not wish to conclude, as Stephen W. Hawking apparently does in *A Brief History of Time*[21], that the scientific world-view is such that a belief in the active presence of God or Being in the cosmos is rather effectively disallowed now have good reasons for arguing and, more importantly, "believing" that this is simply not the case. And yet we will also suggest that if one respects the truths of science and believes that they do provide us with an improved understanding of the conditions for our being and becoming in the vast cosmos, then narrowly anthropomorphic conceptions of the character of Being, or of reality-in-itself, do not seem commensurate with the "vision" of physical reality contained in modern physical theory. What this vision does allow us, in our view, to safely "infer," without, very importantly, being able to "prove," is that the universe is conscious. If one can accept this argument, then the profound sense of alienation that has seemingly been occasioned by the success of classical physics from the eighteenth century to the present could be rather dramatically alleviated. As Abner Shimony[22] has pointed out, the foundations of quantum theory, such as indefiniteness and chance, imply a world-view much more hospitable to resolving the mind-body problem, or the relationship of consciousness to physical reality, than classical metaphysics.

Since our argument is logically consistent and very much in accord with what the "vision" of the cosmos in modern physical theory "infers" about reality-in-itself, one need make in our view only a very small leap of faith to come to the conclusion that the universe is conscious. Yet it is also quite clear that one is not required to make this leap for the same reason that one is "free" to make it—speculations regarding the nature of the undivided wholeness of the cosmos, or reality-in-itself, can be neither proven or disproven by scientific theory and experiment. For those who now have the impression that this discussion is focused exclusively on ontological or metaphysical concerns, we should note that this is not, in fact, the case. We will also advance the hypothesis that the epistemological situation we are obliged to confront in a quantum mechanical universe, in which non-locality must now be viewed as a fundamental fact of nature, provides a new basis for understanding the ability of the human brain to construct symbol systems, or symbolic representations of reality. Drawing extensively on Niels Bohr's definition of the logical framework of complementarity, which we regard as fundamental to understanding the actual character of physical reality in a quantum mechanical universe, we will advance and attempt to support the view that complementarity is the most fundamental dynamic in our conscious constructions of reality in both ordinary and mathematical language systems. If this thesis is correct, it provides a more reasonable and self-consistent explanation than physical scientists have developed thus far as to why the language of mathematics, or the language of mathematical physics, is more "privileged" in its ability to uncover the dynamics of physical reality than is ordinary language. And it could also relieve much of the obvious "angst" that has apparently been occasioned by the rather widespread conviction among humanists and social scientists that all of us are locked, as Nietzsche put it, in the "prison house" of our linguistically-based constructions of reality with no real or necessary connection between subjective reality and external reality.

The most radical hypothesis advanced here is, however, more narrowly scientific. That hypothesis is that since complementarity has been a primary feature in every physical theory advanced in mathematical physics beginning with the special theory of relativity in 1905, and since complementarity can also be shown to be an emergent property or dynamic in the life of the evolving universe at increasingly larger scales and times, then future advances in physical theory in cosmology, or in the study of the origins and evolution of the entire universe, will also feature complementary constructs. In this same discussion, we also suggest that present limits of observation in the study of the large-scale structure of the universe appear to be providing additional evidence that the entire universe is a quantum system, and that cosmologists and astrophysicists may have to invoke complementarity in resolving some seemingly irresolvable problems associated with the most widely accepted model for explaining the origins and evolution of the cosmos—the big-bang model with inflation.

Since this book is, as we have noted, designed to be read by non-physicists, it is likely to be criticized by some physicists and by some experts in the history and philosophy of science as well. Some of this criticism will undoubtedly come from theoretical physicists who will point out that many of the complexities and nuances of our present understanding of the character of physical reality cannot be communicated in a discussion written for the general reader, and they will, to some extent, be quite correct. And similarly valid criticisms are likely to be made by experts in the history and philosophy of science. Although this is simply part of price paid when any expert attempts to translate expert knowledge into a discussion for the general reader, we are quite willing to pay it for the reason indicated earlier. What we confront in the vision of physical reality presently disclosed in modern physical theory should concern, in our view, anyone who has an interest in the implications of this vision for human life and thought. And these implications are, as we hope to demonstrate, sufficiently general so that they can be understood and appreciated by virtually anyone who chooses to direct his or her attention to them.

We should also make it clear at the outset that we will not merely be challenging the views of some humanists-social scientists about the truths of science. Much of what we claim here about the character of physical reality as it has been disclosed in modern physical theory, and particularly the epistemological situation that results, might be viewed as quite heretical by some physical scientists. The simple explanation for why this is likely to be the case is that there is at the moment no "universally" held view, for reasons we will explore, of the actual character of physical reality or of the epistemological implications of quantum physics. It would, of course, be arrogant beyond belief if we were to assume that we have resolved all the relevant dilemmas and provided the basis for such a view. And it would be equally foolish to assume that future progress in physical theory cannot or will not lead to revisions or even refutations of many of the conclusions we have drawn. At the same time we are convinced that the view of physical reality advanced here is quite consistent with the totality of knowledge in mathematical physics, and that our proposed resolution of epistemological dilemmas is very much in accord with this knowledge.

1
Leaving the Realm of the Visualizable: Waves, Quanta, and the Rise of Quantum Theory

Some physicists would prefer to come back to the idea of an objective real world whose smallest parts exist objectively in the same sense as stones or trees exist independently of whether we observe them. That, however, is impossible.
—*Werner Heisenberg*

The Two Clouds of Lord Kelvin

Toward the end of the nineteenth century, Lord Kelvin, one of the best known and most respected physicists at that time, commented that "only two small clouds" remained on the horizon of knowledge in physics. In other words, there were, in Kelvin's view, only two sources of confusion in our otherwise complete understanding of material reality—the results of the Michelson-Morley experiment, which failed to detect the existence of a hypothetical substance called the ether, and the inability of electromagnetic theory to predict the distribution of radiant energy at different frequencies emitted by an idealized "radiator" called the black body. These problems seemed so "small" that some established physicists were encouraging those contemplating graduate study in physics to select other fields of scientific study where there was better opportunity to make original contributions to scientific knowledge. What Lord Kelvin could not have anticipated was that efforts to resolve these two anomalies would lead to relativity theory and quantum theory, or to what came to be called the "new" physics.

What is most intriguing about Kelvin's metaphor for our purposes is that it is visual. We "see," it implies, physical reality through physical theory, and the character of that which is seen is analogous to a physical horizon which is uniformly bright and clear. Obstacles to this seeing, the "two clouds," are likened to visual impediments which will disappear when better theory allows us to see through, or beyond, them to the luminous essence which explains and eliminates them. One reason that the use of such a metaphor would have seemed quite natural and appropriate to Kelvin is that the objects of study in classical physics, like planets, containers with gases, wires, magnets, etc., were perfectly visualizable. His primary motive for metaphor can be better understood, however, in terms of some assumptions about the relationship between the observer and the observed system, and the ability of physical theory to mediate this relationship.

Observed systems in classical physics were understood as separate and distinct from the mind that investigates them, and physical theory was assumed to bridge the gap between these two domains of reality with ultimate completeness and certainty. The measuring instruments were simply more refined means of gathering sensory input, and the expectation was that carefully controlled experiments would, in principle if not always in fact, confirm a one-to-one correspondence between every element of the physical theory and the observed reality. If Lord Kelvin had been correct in assuming that the small clouds would be eliminated through refinement of existing theories, namely Newtonian mechanics and Maxwell's electromagnetic theory, the classical vision would have appeared utterly complete. There would have been no need for a "new" physics, and no need to question classical assumptions about the relationship between physical theory and physical reality.

It is also ironic that light was the primary object of study in the new theories that would displace classical physics. Light in western literature, theology, and philosophy appears rather consistently as the symbol for transcendent, immaterial, and immutable forms separate from the realm of sensible objects and movements. Attempts to describe occasions during which those forms and ideas appear known or "revealed" also consistently invoke light as that aspect of nature most closely associated with ultimate truths. When Alexander Pope in the eighteenth century penned the line, "God said, Let Newton be! and all was Light," he anticipated no ambiguity in the minds of his readers—there was now, assumed Pope, a new class of ultimate truths, physical law and theory, which had been revealed to man in the person of Newton. The irony is that the study of the phenomenon of light in the twentieth century leads to a vision of physical reality that is not visualizable, or which cannot be constructed in terms of our normative seeing in everyday experience.

That classical physics presents us with a picture of physical reality that is visualizable, and that this physics became the foundation in the new physics for a more complete description of physical reality that is unvisualizable, is well known. In fact, most of the *Weltanschauung* theorists in the philosophy of science whose positions were briefly outlined in the introduction have appealed to this movement from the visualizable to the unvisualizable in modern physics to bolster their thesis that the progress of physics is "driven," to a greater or lesser extent, by linguistically-based and culturally derived and determined ideals of order. Although there is, as we will illustrate in our brief accounts of the history of physical theories, some validity to this assumption, the notion that the linguistically-based world-view of the physicist is anything like a fundamental causal factor in the history of physical theories is anything but the case.

Our primary ambition here is to seek to provide a framework within which members of C.P. Snow's two cultures of scientists-engineers and humanists-social scientists can engage one another in a dialogue about the implications of the vision of physical reality provided by modern physical theories in ways that could, in our view, lead to an improved understanding of the conditions for human survival, and also serve to alleviate much of the apparent "angst" that has been occasioned by our previous scientific understanding of the character of physical reality. In an effort to accomplish these admittedly ambitious goals, we will try to disclose and undermine some biases held by members of each culture regarding the special character of the truths of science, and the manner in which these truths originate and evolve.

Part of this effort will be to demonstrate to humanists and social scientists that the progress of physics, contrary to what philosophers of science like Kuhn and Hanson have on occasion suggested, is a continuous and context-driven way of knowing which imposes an increasing number of constraints on what can be viewed as a legitimate concept, problem, or hypothesis. Although classical physics was, for example, premised upon the linguistically-based and culturally derived notion that the actual character of physical reality was visualizable, and a number of extra-scientific or metaphysical assumptions were incorporated into this physics, it was the continued refinement and revision of classical mathematical theory that allowed physics to move rather "continuously" toward the recognition that an improved understanding of this reality was not visualizable. It is "because" the progress of mathematical physics imposes increasingly tighter constraints on what can properly be taken as a description of physical reality that physicists have been led to the recognition that the fundamental character of physical reality is not visualizable. When this progress leads, as is quite usually the case in the modern period, to radically new and counterintuitive results, ideals of order in our linguistically-based symbolic constructions of reality may, in the face of these results, serve to create no small measure of cognitive dissonance in the minds of physicists. Yet what has always triumphed in the history of physical theories is the mathematical theory which proves itself under controlled and repeatable scientific experiments more capable of coordinating greater ranges of experience with physical reality.

This does not mean, however, for reasons developed by the philosopher of science Karl Popper, that a mathematical theory can ever be assumed to attain the status of that which can "never" be contradicted. Even if the results of experiments agree after endless trials with the predictions of a theory, one can never be "completely" assured that the results obtained in subsequent experiments will not contradict the theory. All theories, as Popper suggests, must remain provisional, and all that is required to disprove a theory is a single observation under controlled and repeatable experimental conditions that is not in agreement with it. Also a good theory, as Popper has argued, is characterized by making a number of predictions which, in principle at least, could be disproved or "falsified" with improved experimental techniques and/or new and better theory.

Yet Popper's understanding of the character of physical theory is not, for the physicist at least, grounds for believing that we should abandon the view that physics is a cumulative way of knowing that leads to an increasingly more accurate and complete description of the actual character of physical reality. Popper's commentary on the provisional character of physical theory as well as his notion that "falsifiability" is a criterion for any good theory are viewed by physicists as merely essential aspects of the "process" that allows for this "progress." More important, this progress typically occurs when anomalies are discovered within the context of existing mathematical theories, or in terms of the formal structure of the theories themselves. It is normally the discovery of such anomalies, as we will attempt to demonstrate, that leads to a new class of experiments which extend, refine, or displace previous mathematical theories with improved mathematical theories.

Yet this progress has also recently disclosed, we believe, that one of the fundamental biases of scientists-engineers regarding the epistemological authority of scientific truths and the range of their applications requires some fundamental revisions. The movement into the realm of the unvisualizable has served to disclose the existence of what we will view here as some pre-scientific assumptions about the relationship between physical theory and physical reality that can be traced historically to linguistically-based assumptions in the western metaphysical tradition. Although physicists have been obliged to contend with this prospect since the advent of quantum mechanics in the 1920s and 30s, the full implications have been rather successfully avoided within the community of physicists. The primary mode of avoidance has been to assume that the "strange" and seemingly "philosophical" problems encountered in quantum physics might be eliminated by advances in physical theory. Failing that, some physicists have argued that classical assumptions can still be preserved in our dealings with macro-level phenomena, and, therefore, that the implications of quantum physics can be relegated to the "special" case of dealing with micro-level phenomena.

Although this "wait" with the expectation that we will come to "see" physical reality in the old terms meets the criterion of reasonableness in physical science, there are no indications in our view that the strategy can or will work. The quantum mechanical description of nature has been and continues to be enormously successful. Efforts to evolve a consistent theory for the history of the cosmos, sometimes called the Theory of Everything, or TOE for short, are all premised on quantum mechanics, and any anomalies between quantum theory and other current theories, like relativity, will probably be eliminated with further extensions of quantum theory. Also any alleged distinctions between micro and macro physics, which would allow the implications of quantum theory to be divorced from our dealings in physical theory with macro-level processes, have now begun to take on the appearance of unsubstantiated and unsubstantiatable rationalizations. The scientific evidence suggests, rather overwhelmingly, that we live in a quantum mechanical universe, and that we cannot, therefore, ignore the implications of quantum mechanics in our understanding of the character of all physical theories.

For those who might argue that the quantum mechanical description of nature will over time be utterly displaced by some other description and, therefore, that the implications of this description should not be taken too seriously at this point in time, we concede that this is, in principle at least, a possibility. Yet the continuous progress of physics as well as the range of application of quantum physics both suggest that the more likely and reasonable scenario by far is continued refinement and extension of this existing theory. That these refinements and extensions might lead to some radically different implications about the character of a quantum mechanical universe is, of course, another prospect that cannot be disallowed. All we can say in response to this prospect is that we are convinced that most of the basic features of the universe appear to have been disclosed, and continued extensions and refinements in quantum physics are not likely to fundamentally alter present assumptions about the actual character of physical reality.

Our conviction that we live in a quantum mechanical universe has in our view been massively reinforced of late by a theorem published in 1965, called Bell's theorem, which served to isolate the central questions in the debate over the relationship between physical theory and physical reality in ways that allowed them to be answered in actual experiments. We should also note here that the progress that has been made with the advent of this theorem and the creation of these experiments fits rather nicely into Popper's understanding of the manner in which scientific progress occurs. Bell's theorem originated because he noticed some anomalies in existing theories which allowed him to frame mathematical questions which could serve in actual experiments to "falsify" fundamental aspects of one or the other of the two classes of theories within the domain of physical reality dealt with in the experiments. The experiments made possible by the theorem would serve to falsify the view of the relationship between physical theory and physical reality featured in classical physics or the view of that relationship featured in quantum physics. According to the theorem, both could not "in principle" be correct, and the intent in the experiments themselves was to establish which of the two competing views was correct. Part of what makes these developments dramatic in the history of scientific thought is that they oblige the community of physicists to assess which view of this fundamental relationship is correct within the conduct of science, or in terms of results provided by carefully controlled and repeatable scientific experiments. What is most revealing here, as we shall see, is that many theoretical physicists have made what we will characterize as "metaphysical leaps" in an apparent effort to "save" classical assumptions. Since an appeal to metaphysics is a tendency which physicists have presumably guarded against and expressly forbidden since the early nineteenth century, one wonders why some of the best minds in physics would even flirt with this prospect.

We have long been aware that metaphysics played a large role in the creation of classical physics. The architects of classical physics explicitly appealed to metaphysics with the assumption that doing mathematical physics was a form of communion with immaterial mathematical and geometrical forms resident in the mind of God. Thus the forms, particularly the laws of nature, were presumed to exist in a realm which "transcended," or existed in a realm separate from, that of material reality. As we will discuss in more detail later, metaphysical presuppositions consistent with the "ideals of order" resident in the linguistically-based and culturally derived world-view of the architects of classical physics are also apparent in the classical conceptions of both the character of space and time and the nature of atoms as fundamental building blocks in physical reality. Most trained physicists are not normally made aware of the obvious role played by metaphysics in the creation of classical physics due to the triumph of a view of the character of scientific truths that began to take shape in the eighteenth century. In this century, as Ivor Leclerc explains, "mathematical physics was increasingly regarded as an autonomous science, and any overt appeal to metaphysics as ad hoc and unnecessary."[1] The enormous success of classical physics eventually encouraged a number of nineteenth century physicists and mathematicians to conclude that "concepts like force, mass, motion, cause and laws exist only as 'quantities,' and that any concerns about the 'nature of' or the 'source of' phenomena should be eliminated."[2] This view, which came to be known as positivism, stipulates that "true, genuine and certain knowledge in physics is revealed in the mathematical description, and that all metaphysical concerns should be excluded in both principle and in practice."[3] One of the primary effects of the success of positivism in our view was that it allowed physicists to do physics without any awareness that they were operating on "hidden" metaphysical presuppositions. The simple explanation as to why this could be the case is that experimental conditions and results in classical physics appeared to unambiguously confirm the presuppositions. Those conditions and results provided no reason to doubt either that the observer and the observed system were separate and distinct, or that a one-to-one correspondence between every element of the physical theory and the physical reality actually existed.

Thus metaphysical presuppositions continued in our view to operate in physics not because physicists were displaying an attachment to metaphysics, or unwilling to divorce themselves from metaphysics in the conduct of physics. They survived because the experimental conditions and results appeared to confirm their correctness in the conduct of physics. The interesting result was that physicists could practice physics with the full conviction that they were wholly committed to the positivist program and, therefore, with the untarnished belief that their physics had absolutely nothing to do with metaphysics. Even if one was aware that many of the assumptions that appeared self-evident in the actual conduct of physics could be traced to presuppositions in western metaphysical thought, as virtually all of the great modern theoretical physicists have tended to be, there was no suggestion that they should be viewed as arbitrary. It merely appeared as if the wayward movement of intellectual developments within a particular cultural context resulted in the discovery of some "universal" truths which transcended any and all cultural contexts.

Although quantum mechanics from its very inception has threatened to undermine the classical conception of the relationship between the observer and the observed system and the associated one-to-one correspondence between every element of the physical theory and the physical reality, it was possible to assume until recently that both could be fully restored with continued progress in physics. It now appears, however, that Bell's theorem and the experiments testing that theorem have effectively closed the door on that prospect. This door appears to be finally closing because we are required in this new situation to reexamine classical assumptions about the nature of physical reality before the court of last resort in the conduct of physics—the results from controlled and repeatable scientific experiments. Later in this discussion we will make the case that some physicists, in their efforts to "save" classical assumptions about the relationship between physical theory and physical reality, have "unwittingly" appealed to previously "hidden" metaphysical assumptions, or rather to metaphysical assumptions which had appeared "self-evident" in the normal conduct of physics, in their interpretations of the results of experiments testing Bell's theorem.

If our only impulse in this discussion had been to demonstrate that recent developments in physics require some profound revisions in scientific epistemology, or the rules and procedures of doing physics, this book would have taken a very different form. Although a profound revision in the character of scientific truths, given the authority of those truths in the modern period, would obviously be a major intellectual development, the implications of Bell's theorem and the experiments testing that theorem extend themselves well beyond any narrow concerns about the character of the knowledge we call physics. If we live, as all the current scientific evidence suggests, in a quantum mechanical universe, our inability to make a categorical distinction between observer and observed system does something more than force us to make profound revisions in our understanding of the character of scientific truths. This is so because the experiments testing Bell's theorem were also designed to falsify one or the other of two competing views of the character of physical reality in classical physics and quantum physics. As we noted in the introduction, the experiments would show that physical reality was either local, meaning that correlations between space-like separated regions, or regions that are sufficiently distant from one another so that a signal could not travel from one point to another at speeds faster than light, would not occur. Or, the experiments would show that physical reality is non-local, meaning that correlations between events in space-like separated regions would occur.

That the experiments clearly show that this reality is non-local, or that non-locality must now be viewed as a "fact" of nature, has, in our view, large consequences. Most importantly, it suggests that we must profoundly revise our previous understanding of the relationship between part and whole in physical reality, including the part we call "ourselves" and the whole we call the "cosmos." Although every major development in modern physics since Einstein's special theory published in 1905 has "inferred" this alternate understanding, the results of the experiments testing Bell's theorem appear to have taken it out of the realm of inference and firmly planted it in the realm of empirically demonstrated fact. If the current debate over the meaning of the amazing new fact of nature called non-locality resolves itself as we predict it will, it could also lead us to an alternate understanding of the character of human consciousness, and the relationship of this consciousness to the rest of the cosmos.

For those who now have the expectation that this is merely another attempt to draw parallels between modern physics and some established version of metaphysics, which normally requires that one play fast and loose with the implications of modern physical theories, let us emphatically state at the outset that this is not the case. In all our dealings with the history and content of physical theories, as we noted earlier, the commitment is to metaphysical and epistemological realism, and it is this commitment which will finally lead us to demonstrate that modern physics cannot, in principle, be used to support or refute metaphysical assumptions as they have been defined in any mytho-religious heritage. In other words, nature is perfectly silent in the face of our efforts to "legitimate" fundamental metaphysical assumptions about our being and becoming in the vast cosmos. The large paradox, which we will spend some time exploring, is that the inability of physical science to resolve questions concerning the ultimate character of Being frees us in unexpected ways to recognize and pursue, within the context of a scientific world view, a far more profound relationship between our conscious awareness of reality and reality-in-itself. And yet this same freedom also allows us, we should emphasize, to completely ignore or deem irrelevant any such relationship.

One of the obvious reasons that the truths of science tend to have more authority than other truths among educated people is that they are subject to demonstration or "proof" under experimental conditions in spite of their "provisional" character. Yet there is, as we have suggested here, another vital aspect of this process. Science learns which questions to ask in controlled and repeatable experiments because this way of knowing is more cumulative and context driven than any other way of knowing. What this means for our purposes is that the questions asked by Bell in his theorem and answered in experiments can only be fully understood within the context of prior developments in physics. In these terms science is rather unforgiving—one must definitely spend some time becoming familiar with the fundamentals of this music before any sounds become recognizable as tunes. This discussion will, therefore, begin with some "tuning the piano" exercises which examine those aspects of prior developments in physics which are essential to this discussion.

Our first task will be to examine the manner in which the voyage into the realm of the unvisualizable in modern physics served to undermine the classical conception of "seeing" the truths of nature through the lenses of physical theory. This material will also serve as background later in this discussion for appreciating how the visualization problem served to "obscure" or "frustrate" a proper understanding of the implications of the quantum mechanical view of nature. Since our emphasis is upon the scientific issues and problems which cannot be understood without an appeal to the mathematical description of nature, we will appeal throughout to that description. Yet the central issues and problems can also be understood and illustrated with a minimal appeal to scientific formalism, or in a manner in which someone without background in higher mathematics can easily understand.

Since this approach does not do any violence to these issues and problems from the perspective of those who have this background, and yet frames them within an historical and philosophical context that may not be familiar to math and science majors, it should appeal to these readers as well. The difference is that the degree of difficulty presented by the various chapters will probably vary as a function of relative exposure to C.P. Snow's "two cultures." Yet understanding our new situation demands, as we shall see, that we migrate from one of these cultures to the other, and a willingness to do leads, as we will try to demonstrate, to some very satisfying results. For readers who are expert in quantum physics, the material in this chapter as well as in chapters 2 and 3 will probably seem obvious and elementary. And the same could well be the case for many, if not most, experts in the history and/or philosophy of science not only in these chapters but in chapters 4 and 5 as well. We recommend, however, that experts in these areas should at least scan the chapters containing familiar material for the following reason—much of the discussion in chapters 1–5 is designed to provide a conceptual framework within which the more speculative and original material contained in chapters 6–9 can be more easily appreciated and understood.

Light and Relativity Theory

It was, as we noted earlier, the study of light which moved physics out of the realm of the visualizable and into the realm of the unvisualizable. The best known of these experiments is probably that conducted in 1887 by Albert A. Michelson and Edward W. Morley. Their intent was to refine existing theory, in this case Maxwell's electromagnetic theory, and both scientists were terribly disappointed when the effort failed. Light in Maxwell's theory is visualizable as a transverse wave consisting of magnetic and electric fields which are perpendicular to each other, and to the direction of propagation of the wave. This wave theory of light had been established since the early 1800s, and was well supported in experiments on light in which behavior like interference and diffraction had been observed. Interference arises when two waves, like those produced when two stones fall on the surface of a pond, combine to form larger waves when the crests of the two waves coincide, or cancel one another out when the crest of one wave corresponds with the trough of another. Diffraction is a wave property evident when waves bend around obstacles, like when ocean waves go around a wave-breaker in a harbor. Since all previously known wave phenomena propagated through a material medium, it was natural to assume that light, which was presumed to consist of electromagnetic waves, required a material medium through which its vibrant energy could propagate as well. The visualizable material medium whose existence

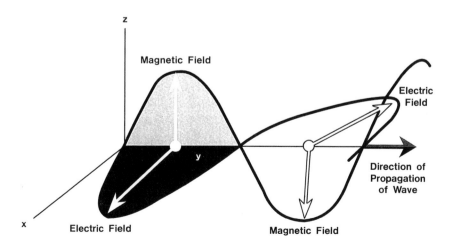

FIGURE 1. Light as an electromagnetic wave.

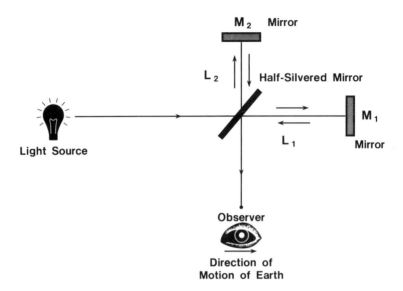

FIGURE 2. The Michelson-Morley experiment.

was implied in the visualizable theory was, however, only a hypothesis, and Michelson and Morley were attempting to prove in experiment that the hypothetical medium, called the ether, was actually there.

Although the ether, to be consistent with classical theory, would have to fill all of space, including the vacuum, and evince the stiffness of a material much stiffer than steel, Michelson and Morley were convinced that it would be found if an appropriate experiment was set up. What is suggested in their conviction that the experimental results would be positive is not naivete, but rather how complete the classical description appeared to physicists at the end of the nineteenth century. In the experiment a new device, called an interferometer, allowed accurate measurement of the speed of a beam of light, and the experiment itself, like most great experiments in physics, was as simple as it was elegant.

An original beam from a light source was split into two beams by means of a half-silvered mirror, and each beam was allowed to travel an equal distance along their respective paths. Assuming that the hypothetical ether was absolutely at rest, the scientists decided to allow one beam of light to move in the direction that the earth moves, and the second beam to move at 90 degrees with respect to the first. The assumption was that the beam moving in the direction of the earth's movement would travel faster as it traveled through the ether, due to the increase in velocity provided by the rotation of the earth. Since that increase in velocity would not be a factor for the beam moving at 90 degrees with respect to the first, the expected result was that the interferometer would show a difference in the velocity of the two beams, and thereby confirm the actual existence of the ether. Yet no difference was found in the actual experiments. This null result carried an implication which seemed as strange to Michelson and Morley as non-locality seems strange in the experiments testing Bell's theorem—the speed of light is constant. Although Einstein's relativity theory had not yet been invented to account for the null result, that theory would eventually explain it. And it is here that physics takes the first great leap out of the realm of the visualizable, and into a quite different conception of the character of physical reality which is unvisualizable.

As we shall see in more detail later, Einstein did not arrive at relativity theory in the effort to account for the unexpected results of the Michelson-Morley experiment. He was seeking to eliminate some asymmetries in mathematical descriptions of the behavior of light, or electromagnetic radiation, in Newtonian mechanics and Maxwell's electromagnetic theory. The Newtonian construct of three-dimensional absolute space existing separately from absolute time implied that one could find a frame of reference absolutely at rest. Newtonian mechanics also implied that if one traveled fast enough, one could catch up with light, and that in this frame of reference the speed of light would be reduced to zero. Einstein's first postulate was that it is impossible to determine absolute motion, or motion that proceeds in a fixed direction at a constant speed. The only way, he reasoned, that we can presume such motion is to compare it with that of other objects. In the absence of such a comparison, which is what "absolute" motion assumes, one can make no assumptions about movement. Thus the assumption that there is an absolute frame of reference in which the speed of light is reducible to zero, which the classical view of space and time as absolute and separable dimensions requires, must be false.

After concluding that there is no absolute frame of reference, or that the laws of physics hold equally well in all frames of reference, Einstein arrived at his second postulate of the absolute constancy of the speed of light for all moving observers. Based on these two postulates, the relativity of motion and the constancy of the speed of light, the entire logical structure of relativity theory followed. As Einstein's friend and colleague Paul Ehrenfest pointed out, however, there is also a third assumption here, which is implied as opposed to stated, which reconciles what would otherwise be a contradiction between the two stated postulates. The contradiction is simply that the assumption that all uniform motions are relative to one another is not consistent with the assumption that the motion of light is absolute. The third assumption, which resolves the contradiction, is that the absolute constancy of the speed of light allows us to understand why the motion of everything else is relative.

Using these postulates, Einstein mathematically deduced the laws that related space and time measurements made by one observer to the same measurements made by another observer moving uniformly relative to the first. Although the French mathematician Poincare had independently discovered the space-time transformation laws in 1905, he saw them as postulates without any apparent physical significance. Since Einstein perceived that the laws did have physical significance, he is recognized as the inventor of relativity. One consequence was that the familiar law of simple addition of velocities does not hold for light, or for speeds close to the speed of light. The reason why these relativistic effects are not obvious in our everyday perception of reality, suggested Einstein, is that light speed is very large compared with ordinary speeds. The highest speed attainable by an intercontinental ballistic missile is, for example, no more than a few kilometers per second, while light travels at 300,000 kilometers per second.

The primary impulse behind the special theory was a larger unification of physical theory which would serve to eliminate mathematical asymmetries apparent in existing theory. There was certainly nothing new here in the notion that frames of reference in conducting experiments are relative—Galileo arrived at that conclusion. What Einstein did, in essence, was extend the so-called Galilean relativity principle from mechanics, where it was known to work, to electromagnetic theory, or the rest of physics as it was then known. What was required to achieve this greater symmetry was to abandon the Newtonian idea of an absolute frame of reference and, along with it, the ether. This led to the conclusion, as Einstein put it, that the "electrodynamic fields are not states of the medium [the ether] and are not bound to any bearer, but they are independent realities which are not reducible to anything else."[4] In a vacuum, light traveled, he concluded, at a constant speed, c, equal to 300,000 km/sec, and thus all frames of reference become relative. There is, therefore, no frame of reference absolutely at rest. This meant that the laws of physics could apply equally well to all frames of reference moving relative to each other.

Einstein also showed that the results of measuring instruments themselves must change from one frame of reference to another. For example, clocks in the two frames of reference would not register the same time. Consequently, two simultaneous events in a moving frame would appear to occur at different times in the unmoving frame. This is precisely what the Lorentz transformation equations, which allow us to coordinate measurements in one frame of reference moving with respect to a second frame, show to be the case. For the observer in the stationary frame, lengths in the moving frame appear contracted along the direction of motion by a factor of $(1 - v^2/c^2)^{1/2}$ where v is the relative speed of the two frames. Masses in the moving frame also appear larger to the stationary frame by the factor $1/(1 - v^2/c^2)^{1/2}$. In the space-time description used to account for the differences in observation between different frames, time is just another coordinate in addition to the three space coordinates forming the four-dimensional space-time continuum. In relativistic physics, transformations between different frames of reference express each coordinate of one frame as a combination of the coordinates of the other frame. A space coordinate, for example, in one frame usually appears as a combination, or mixture, of space and time coordinates in another frame.

What is being illustrated here is that the abandonment of the concept of an absolute frame of reference moves us out of the realm of the visualizable into the realm of the mathematically describable but unvisualizable. Although we can illustrate light speed with visualizable illustrations, like approaching a beam of light in a spacecraft at speeds fractionally close to that of light and imagining that the beam would still be leaving us at its own constant speed, the illustration bears no relation to our direct experience with differences in velocity. It is when we try to image the four-dimensional reality of space-time as it is represented in mathemati-

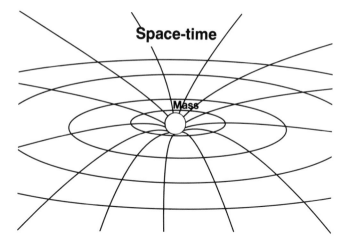

FIGURE 3. Warped space-time around a gravitating mass.

cal theory that we have our first dramatic indication of the future direction of physics. It cannot be done no matter how many helpful diagrams and illustrations we choose to employ.

And yet, as we have also discovered in numerous experiments, the counterintuitive results predicted by the theory of relativity occur in nature. For example, unstable particles, like muons, which travel close to the speed of light and decay into other particles with a well-known half-life, live much longer than their twin particles moving at lower speeds. Einstein was correct—the impression that events can be arranged in a single unique time sequence (past, present, and future), and measured with one universal physical yardstick is due to the fact that the speed of light is so large compared with other speeds that we have the illusion that we "see" an event in the very instant in which it occurs.

In order to illustrate that simultaneity does not hold in all frames of reference, Einstein used a thought experiment featuring the fastest means of travel for human beings at his time—trains. What would happen, he wondered, if we were on a train that actually attained light speed? The answer is that lengths along the direction of motion would become so contracted as to disappear altogether, and clocks would cease to run entirely. Three-dimensional objects would actually appear rotated so that a stationary observer could see the back of a rapidly approaching object. To the moving observer, all objects would appear to be converging on a single blinding point of light in the direction of motion. Yet the train, as Einstein knew very well, could not "in principle" reach light speed. Any configuration or "manifestation" of matter other than massless photons, or light, cannot reach light speed because, in accordance with the equivalence of mass and energy, its mass would have to become infinite, and an infinite amount of energy would be required to get there as well. Here again common-sense, linguistically-based assumptions fail us. When we turn to the mathematics, however, the situation is clear. Light or photons have zero rest mass, and travel exactly at light speed. In accordance with the Lorentz transformations, the factor $(1 - v^2/c^2)^{1/2}$ becomes zero as the relative speed v approaches light speed c.

The special theory of relativity dealt only with "constant," as opposed to "accelerated," motion of the frames of reference. The Lorentz transformations apply to frames moving with uniform motion with respect to each other. The central idea in general relativity theory is that it is impossible to distinguish between the effects of gravity and of nonuniform motion. If we did not know, for example, that we were on an accelerating spaceship, and dropped a cup of coffee, we could not determine whether the mess on the floor was due to the effects of gravity or the accelerated motion. This inability to distinguish between a nonuniform motion, like an acceleration, and gravity is known as the principle of equivalence. In 1915–1916, Einstein extended relativity to account for the more general case of accelerated frames of reference in his general theory of relativity.

Here Einstein posits the laws relating space and time measurements carried out by two observers moving uniformly, as in the example of one observer in an accelerating spaceship and another on earth. Force fields, like gravity, cause space-time, Einstein concluded, to become warped or curved, and hence non-Euclidean in form. In the general theory the motion of material points, including light, is not along straight lines, as in Euclidean space, but along "geodesics" in curved space. The necessity to use curved geometries resulted in yet another higher level of mathematical complexity which Einstein developed based on the work of the great nineteenth-century mathematicians Georg Riemann, Karl Gauss, and Nikolai Lobachevskii. The movement of light along curved spatial geodesics was confirmed in an experiment performed during a total eclipse of the sun by Arthur Eddington in 1919.

Here, as in the special theory, visualization does not work. This is nicely illustrated in the typical visual analogy used to illustrate what spatial geodesics mean. In this analogy we are asked to imagine a hypothetical flatland which, like a tremendous sheet of paper, extends infinitely in all directions. The inhabitants of this flatland, the flatlanders, are not aware of the third dimension. Since the world here is perfectly Euclidean, any measurement of the sum of the angles of triangles in flatland would equal 180 degrees, and any parallel lines, no matter how far extended, would never meet. We are then asked to move our flatlanders to a new land on the surface of a large sphere. Initially, our relocated population would perceive their new world as identical to the old, or as Euclidean and flat. Next we suppose that the flatlanders make a technological breakthrough which allows them to send a kind of laser light along the surface of their new world for thousands of miles. The discovery is then made that if the two beams of light are sent in parallel directions, they come together after traveling a thousand miles.

After at first experiencing utter confusion in the face of these results, the flatlanders eventually realize that their world is non-Euclidean or curved, and invent Riemannian geometry to describe the curved space. The analogy normally concludes with the suggestion that we are the flatlanders, with the difference being that our story takes place in three, rather than two, dimensions in space. Just as the shadow creatures could not visualize the curved two-dimensional surface of their world, so we cannot visualize a three-dimensional curved space. Thus a visual analogy used to illustrate the reality described by the general theory is useful only to the extent that it entices us into an acceptance of the proposition that the reality is totally unvisualizable. Yet here, as in the special theory, there is no ambiguity in the mathematical description of this reality. Although curved geodesics are not any more unphysical than straight lines, visualizing the three spatial dimensions as a "surface" in the higher four-dimensional space-time cannot be done.

The Rise of Quantum Theory

The removal of Kelvin's second small cloud represented the first step in the development of a description of physical reality that is even more unvisualizable than that disclosed by relativity theory—quantum theory. This step was taken by German physicist Max Planck as he addressed the problem of the inability of current theory to explain black body radiation, and the object of study was, once again, the behavior of light. A black body absorbs virtually all radiation that falls on it and emits radiant energy in the most efficient way as a function of its temperature. If you take, for example, a material object, like a metal bar, put it in a dark, light-tight room, and heat it to a high temperature, it will produce a distribution of radiant energy with wavelengths or colors that can be measured. If we make precise measurements of this radiation as the metal bar achieves higher temperatures and changes from dark red to white hot, a black body radiation curve can be obtained which has a bell-shaped appearance.

The assumption in physics at the end of the nineteenth century was that a black body radiates when the multitude of tiny charged particles inside it emit energy as they rapidly vibrate. The emission from these "alleged" vibrating electrical charges was described by electromagnetic theory. The "cloud" here was that when the emission from all the vibrating charges was summed in accordance with electromagnetic theory, it predicted infinities as the frequency of light increased, and this was clearly not in agreement with the observed bell-shaped behavior of black body intensity.

Planck, working with experimental results developed by a team of experimental physicists at the Physikalisch-Technische Reichsanstalt in Berlin, had tackled this problem before Kelvin's two cloud address, and, like Michelson and Morley before him, was not comfortable with the results. After failing to reconcile the results with existing theory, Planck, out of what he later described as "an act of sheer desperation," made a large intuitive leap. His great insight was that the vibrating charges do not, as classical theory said they should, radiate light with all possible values of energy continuously. Supposing that the material of the black body consisted of "vibrating oscillators," which would later be understood as subatomic events, he concluded that the energy exchange with the black body radiation might be discrete or quantized. Following this hunch, he then viewed the energy radiated by a vibrating charge as an integral multiple of a certain unit of energy for that oscillator. What he found was that the minimum unit of energy is proportional to the frequency

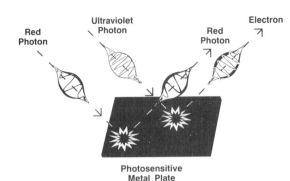

FIGURE 4. The photoelectric effect: A photon of low energy (red) cannot eject an electron but a photon of high energy (ultraviolet) can.

of the oscillator. Working with this proportionality constant and calculating its value based on the careful data supplied by the experimental physicists, Planck solved the black body radiation problem. Although Planck could not have realized it at the time and would in ways live to regret it, his announcement of the explanation of black body radiation on December 14, 1900, was the birthday of quantum physics. Planck's new constant, known as the quantum of action, would later apply to all microscopic phenomena. The fact that the constant, h, is, like the speed of light, a universal constant would later serve to explain the strangeness of the new and unseen world of the quantum, and lead inexorably to a new way of viewing the universe.

The next major breakthrough was made by the physicist who would eventually challenge the emergent new quantum paradigm with the greatest precision and fervor. In the same year (1905) that the special theory appeared, Einstein published two other seminal papers that also laid foundations for the revolution in progress. One, which we will discuss in a moment, was on the so-called Brownian movement, and the other was on the photoelectric effect. In the paper on the photoelectric effect, Einstein challenged once again what had previously appeared in theory and experiment as obvious, and the object of study was, once again, light. The effect itself was a by-product of Heinrich Hertz's experiments, which at the time were widely viewed as having provided conclusive evidence that Maxwell's electromagnetic theory of light was valid. When Einstein explained the photoelectric effect, he showed precisely the opposite result—the inadequacy of classical notions to account for this phenomenon.

The photoelectric effect is witnessed when light with a frequency above a certain value falls on a photosensitive metal plate and ejects electrons. A photosensitive plate is one of two metal plates connected to ends of a battery, and placed inside a vacuum tube. If the plate is connected to the negative end of the battery, light from a light source falling on the plate can cause electrons to be ejected from the negative end, and these electrons travel through the vacuum tube to the positive end, producing a flowing current.

In classical physics the amplitude, or height, of any wave, including electromagnetic waves, describes the energy contained in the wave. The problem which Einstein sought to resolve can be thought about using water waves as an analogy.

Large water waves, like ocean waves, have large height or amplitude, carry large amounts of energy, and are capable of moving many pebbles on a beach. Since the brightness of a light source is proportional to the amplitude of the electromagnetic field squared, it was assumed that a bright source of light should eject lots of electrons, while a weak source of light should eject few electrons. In other words the more powerful wave, the bright light, should move more pebbles, or electrons, on this imaginary beach. The problem was that a very weak source of ultraviolet light was observed to be capable of ejecting electrons, while a very bright source of lower frequency light, like red light, did not. It was as if the short, choppy waves from the ultraviolet source could move pebbles, or electrons, on this imaginary beach, while the large waves from the red light source could not move any at all.

Einstein's explanation for these strange results was as simple as it was bold. He argued that the energy of light is not distributed, as classical physics supposed, evenly over the wave, but is rather concentrated in small, discrete bundles. Rather than view light as waves, Einstein conceived of light as bundles, or "quanta," of energy. The reason that ultraviolet light ejects electrons and red light does not, concluded Einstein, is that the energy of these quanta is proportional to the frequency of light, or to its wavelength. Using this quantum picture, it is the energy of the individual quanta, rather than the brightness of the light source, that matters. Viewing the situation in these terms, individual red photons do not have sufficient energy to knock an electron out of the metal while individual ultraviolet photons have sufficient energy. When Einstein computed the constant of proportionality between energy and the frequency of the light, or photons, he discovered, much to his amazement, that it was equal to Planck's constant, h.

The Atomic Theory

The next step on the road to quantum theory was made by a Danish physicist from whom we will hear a great deal more later in this discussion—Niels Bohr. Developed partly as a result of the work which he did with Ernest Rutherford in Manchester, Bohr provided, in a series of papers published in 1913, a new model for the structure of atoms. At the time the physical dynamics of the unvisualizable realm of subatomic events were extremely vague, and the visualizable picture was drawn from analogies with macro-level phenomena. The discovery of the element polonium, by Pierre and Marie Curie in 1898, had previously suggested that atoms were composite structures that transformed themselves into other structures as a result of radioactivity. Einstein's paper on Brownian motion published in 1905 also enticed physicists to conceive of atoms as something more than a philosophical construct in the manner of the ancient Greeks. The motion is called Brownian after the British botanist Thomas Brown, who discovered in 1827 that when a pollen grain floating on a drop of water is examined under a microscope, it appears to move randomly. Einstein showed that this motion obeys a statistical law, and the pattern of motion can be explained if we assume that objects, like pollen grains, are moving about as they collide at the microscopic level with tiny molecules of the water.

Although Einstein did suggest that the molecules and the atoms which constitute them were real in that their behavior had concrete effects on the macro level, nothing of substance was known at the time about the internal structure of atoms. We should also note that when Einstein used the word "statistical" in his paper on Brownian motion, he was not using that term as it would later be used in quantum mechanics. He was using it in the classical sense, or in the manner defined in the statistical theories of Maxwell and Ludwig Boltzmann. The word as Einstein used it refers to the average behavior of a large number of particles, like those in gases, which we must predict statistically because the behavior of the individual particles is not directly accessible for measurement. Thus Einstein's presumption in the paper on Brownian motion was that the behavior of the individual particles would still follow the laws of classical dynamics, and that a one-to-one correspondence between every element of the physical theory and the physical reality was not threatened in the least.

The suggestion that the world of the atom had a structure enticed Rutherford to conduct a series of experiments in which positively charged "alpha" particles, later understood to be the nuclei of helium atoms, were emitted from radioactive substances, and fired at a very thin sheet of gold foil. If there was nothing to impede the motion of the particles, they should travel on a straight line and collide with a screen of zinc sulfide where a tiny point of light, or scintillation, would record the impact. In this experiment most of these particles were observed to be slightly deflected from their straight line path. Other alpha particles, however, were observed to be deflected backwards towards the direction from which they came. Based on an estimate of the number of alpha particles emitted by a gram of radium in one second, Rutherford was able to arrive at a more refined picture of the internal structure of the atom.

The existing model, invented by the discoverer of the electron, J.J. Thomson, presumed that the positive charge was distributed over the entire space of the atom. The observed behavior of the alpha particles suggested, however, that some particles, those which were deflected backwards, were encountering a highly concentrated positive charge, while most of them traveled through the space of the atoms as if this space were empty. Rutherford explained the results in terms of a picture of the atom as being composed primarily of vast regions of space in which the negatively charged particles, electrons, move around a positively charged nucleus which contains by far the greatest part of the mass of the atom.

Forced to appeal to macro-level analogies to visualize this unvisualizable structure, Rutherford termed the model "planetary." Soon it was discovered that there is practically no similarity between the structure or behavior of macro and micro worlds. Our sun, the most massive body in the solar system, bears no resemblance to the tiny nucleus, and the relative distances between electrons and nucleus, as compared to the size of the nucleus, are much greater than the relative distances between planets and the sun, as compared to the size of the sun. Also, the dynamical properties of atoms would shortly obviate any comparisons whatsoever

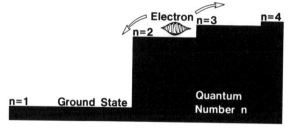

FIGURE 5. The energy levels in the Bohr atom can be visualized as a set of steps of different heights. The electron, visualized here as a wave packet, is always constrained to be found on one of the steps.

with any macro-level phenomena. Imagine Earth undergoing a quantum transition and instantaneously appearing in the orbit of Mars, and we begin to appreciate how inappropriate macro-level analogies would soon become.

Although Bohr had no choice but to use macro-level analogies, he was the first to suggest that the orbits of electrons were quantized. Bohr's model was semi-classical in that it incorporated ideas from classical celestial mechanics about orbiting masses. The problem he was seeking to resolve had to do with the spectral lines of hydrogen, which showed electrons occupying specific orbits at specific distances from the nucleus with no in-between orbits. Spectral lines are produced when light from a bright source containing a gas, like hydrogen, is dispersed through a prism. The pattern of the spectral lines is unique for each element. The study of the spectral lines of hydrogen suggested that the electrons somehow "jump" between the specific orbits, and as they do so they appear to absorb or emit energy in the form of light or photons. What, wondered Bohr, was the connection? Bohr discovered that if you use Planck's constant in combination with the known mass and charge of the electron, the approximate size of the hydrogen atom could be derived.

Assuming then that a jumping electron absorbs or emits energy in units of Planck's constant in accordance with the formula Einstein used to explain the photoelectric effect, Bohr was able to find correlations with the specific spectral

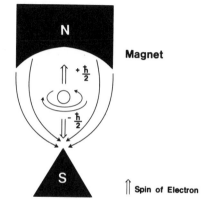

FIGURE 6. Quantization of spin: Along a given direction in space, the measured spin of an electron can only have two values.

lines for hydrogen. More important, the model also served to explain why the electron does not, as electromagnetic theory says it should, radiate its energy quickly away and collapse into the nucleus. Bohr reasoned that this does not occur because the orbits are quantized—electrons absorb and emit energy corresponding to the specific orbits. Their lowest energy state, or lowest orbit, is the "ground state."

The central problem with Bohr's model from the perspective of classical theory was pointed out by Rutherford shortly before the first of the papers describing the model was published. "There appears to me," Rutherford wrote in a letter to Bohr, "one grave problem in your hypotheses which I have no doubt you fully realize, namely, how does an electron decide what frequency it is going to vibrate at when it passes from one stationary state to another? It seems to me that you would have to assume that the electron knows beforehand where it is going to stop."[5] Viewing the electron as atomic in the Greek sense, or as a point-like object that moves, then there is, indeed, cause to wonder, in the absence of a mechanistic explanation, how this object instantaneously "jumps" from one shell or orbit to another. It was essentially efforts to answer this question that led to the development of a complete quantum theory.

The effect of Bohr's model was to raise more questions than it answered, like the inability of the model to predict the energy levels of atoms more complicated than hydrogen. Although the model allowed Bohr to suggest that we can explain the periodic table of the elements by assuming a maximum number of electrons are found in each shell, he was not able to provide any mathematically acceptable explanation for the hypothesis. That explanation was provided in 1925 by Wolfgang Pauli, known throughout his career for his extraordinary talents as a mathematician. Bohr had used three quantum numbers in his model—Planck's constant, mass, and charge. Pauli added a fourth, described as "spin," which was initially represented with the macro-level analogy of a spinning ball on a pool table. Rather predictably the analogy does not work. Whereas a classical spin can point in any direction, a quantum mechanical spin points either up or down along the axis of measurement.

In fact, in total contrast to the classical notion of a spinning ball, if no axis of measurement is defined, we cannot even speak of the spin of the particle. For this reason, spin in quantum theory is described as a double valued number which, in the case of the electron, has a value of one half $\hbar$ (where $\hbar$ is Planck's constant divided by 2π). When Pauli added this fourth quantum number, he found a correspondence between the number of electrons in each full shell of atoms, and the new set of quantum numbers describing the shell. This became the basis for what we now call the Pauli exclusion principle. The principle is simple and yet quite startling—two electrons cannot have all their quantum numbers the same. This presents us with a sameness and difference paradox. Although all the properties of electrons are known, no two actual electrons are identical in the sense of having the same quantum numbers. The exclusion principle explains mathematically why there are a maximum number of electrons in the shell of any given atom.

If the shell is full, adding another electron would be excluded because this would result in two electrons in the shell having the same quantum numbers.

Pauli also discovered that if all the other quantum numbers of two neighboring electrons are equal, their spins must be opposite. This led to a distinction between the properties of quantum particles based on spin. Particles with a half integer spin (1/2, 3/2, etc., of the fundamental unit of $\hbar/2\pi$) are called "fermions," after the physicist Enrico Fermi, and obey the exclusion principle. Particles with full integer spin (1, 2, 3, etc., of the unit $\hbar/2\pi$) are called bosons, and do not obey the exclusion principle. Although this may sound a bit esoteric, the distinction between fermions and bosons is, when we think about it, quite fortunate. Our very existence depends on this difference.

The electron, proton, and neutron are fermions, and obey the exclusion principle. If they did not, the electrons of all elements would exist only at the ground state, with no chemical affinity between them, and there would be no compounds. All elements would be similar to the hydrogen atom. Structures like crystals or DNA would not exist, and the only structures that could exist would be spheres held together by gravity. Photons are "bosons," after the physicist Bose who, along with Einstein, studied their statistics, and do not obey the exclusion principle. If photons could not crowd together in the energy of light, the light energy which fuels the quantum mechanical processes that lead to the evolution of chemical structures, including what we call life, would not exist. We coordinate our experience with fermions and bosons with fundamentally different statistics. The point, once again, is that the relationship between the unvisualizable theory, which better explains events in nature, and the mathematical complexity of that theory is intimate and ongoing.

Waves as Particles and Particles as Waves

The next development in the progress toward quantum theory was based upon the results of experiments conducted by Arthur Compton using X-rays, published in 1923. Einstein had earlier discovered that photons possess a momentum, p, related to their wavelength, λ, by the simple relation $p = \hbar/\lambda$. Compton found that in a collision of an X-ray photon with an electron, the total momentum of the system is conserved, and the wavelength of light changes appropriately. The results suggested that the photons were behaving like particles. If light had particle properties, when it was previously conceived as a wave, then perhaps the electron, previously conceived as a particle, had wave properties as well. A French doctoral student in physics, Louis de Broglie, suggested in his thesis that the same formula found by Einstein to hold for photons, and applied by Compton to the collisions of photons with electrons, might apply to all known particles. This would mean that an electron should also possess a wavelength, λ, related to its momentum, p, by the same formula $p = \hbar/\lambda$. This came to be known as the de Broglie wavelength.

Appealing once again to macro-level analogies, the existence of the so-called *matter waves* was demonstrated in experiments involving the scattering of

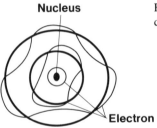

Nucleus

FIGURE 7. De Broglie's explanation of why electrons are confined to specific orbits in terms of standing waves.

Electron

electrons off crystals. Electrons showed interference patterns indicative of wave properties. The consensus would eventually become that particles possess wave-like properties in the same way that light possesses particle-like properties. Thus de Broglie's hunch led to a large and unexpected unification. It allowed an explanation for the previously unexplained assertion in Bohr's model that an electron is confined to specific orbits. An electron, concluded de Broglie, is confined to orbits in terms of integer numbers of waves. De Broglie's thesis was brought to the attention of Einstein, who then brought it to the attention of Erwin Schrödinger, a professor in Zürich. Drawing on his own classical understanding of wave phenomena, Schrödinger proceeded to develop wave mechanics (1925). The nineteenth century physicist William Hamilton had created a series of equations describing the geometrical particle-like and wave-like properties of light. Drawing on Hamilton's equations, Schrödinger assumed the reality of "matter waves" which behave in accordance with a wave function ψ. This led to the formulation of wave mechanics.

We now know that this theory is merely one aspect of a complete quantum theory, and that the "matter waves" are not "real" as water waves are real. We would also eventually learn that our impulse to visualize the unvisualizable realm of micro-level processes and, consequently, to confer our normative macro-level sense of reality on the dynamics of the micro level, only serves to distort the scientific reality. But that impulse, coupled with the success of wave mechanics in the laboratory, would frustrate efforts to fundamentally understand quantum reality until quite recently.

The first insight that would open the door to an improved understanding came from Max Born in 1926, and it was not well received by the majority of physicists at the time. Born realized that the wave function itself, ψ, cannot be observed. The square of the wave function, concluded Born, gives us the "probability" of finding a particle within a region. But we cannot precisely predict where that particle will be found. Born was defining the word "probability" in the sense that we would eventually be obliged to use it, as an inherent aspect of measurement of all quantum mechanical events. He was not using the word in the classical sense, meaning a convenient way of assessing the behavior of the system on the presumption that an actual one-to-one correspondence between theory and reality could theoretically be proven under improved experimental conditions. Since this was the first direct assault on classical assumptions about the relationship between physical theory and physical reality, it was, understandably, viewed as heretical by most physicists.

Born's recipe features an aspect of quantum theory that not only cannot be found in classical physics, but which is also not in accord with any familiar feature of the everyday world. Although the recipe is simple mathematically, the reality it describes is totally unvisualizable. The wave function, ψ, is unobservable, but the square of the wave function gives us the probability of finding the particle within a particular region of space with certain properties. To be more precise, the wave function, ψ, is a complex function, i.e., $\psi = a + bi$, where a and b are real functions and $i = (-1)^{1/2}$. One must, therefore, compute its absolute value or amplitude by squaring the wave function, $|\psi|^2$. It is as if the wave function, ψ, defines the possibilities, and the experimental results are only predictable in terms of probabilities, i.e., probability = $|possibility|^2$.

The wave function provides a complete description of the quantum particle or system, and in this sense at least, wave mechanics is a "complete" theory. Yet in practice, or in actual experiments, the theory describes only probabilities of events happening, as opposed to specific events. The completeness of quantum theory is observable only in the sense that in a given situation only one possibility is materialized in accordance with a probability that can be calculated. The specific event cannot be predicted; we can predict only the "probability" that it may happen. Einstein characterized the strangeness of this situation from a classical point of view by referring to the wave function as a "ghost field." Rather than represent a "real" matter wave, the wave function describes, suggests Einstein, only a wavy, probabilistic "reality." Although this situation may seem simple enough mathematically, the real existence of wave and particle aspects of reality is much stranger in practice than it seems in principle. And attempts to explore the implications in a quantum mechanical universe have led, as we shall see, to a number of theories which, to the uninitiated, could seem utterly bizarre.

In the same year (1925) that Schrödinger was developing wave mechanics, Werner Heisenberg, Max Born, and Pascual Jordan were constructing an alternate set of rules for calculating the frequencies and intensities of spectral lines. Operating on the assumption that science can only deal in quantities that are measurable in experiments, their focus was on the particle aspect. The result was matrix mechanics. Matrices involve calculations with a quite curious property. When two matrices are multiplied, the answer that we get depends on the order of their multiplication. In other words, for matrices 2×3 would not be equal to 3×2, or, in the language of algebra, $a \times b \neq b \times a$. The word "matrix" is used here because in the Heisenberg-Born-Jordan formulation of quantum theory the alternate set of rules applied to organizing data into mathematical tables, or matrices. These tables were used to calculate probabilities associated with initial conditions that could be applied in the analysis of observables. As Heisenberg would reflect on the situation later, we have now arrived at the point where we must "abandon all attempts to construct perceptual models of atomic process."[6]

It is also significant that the point at which we fully enter via mathematical theory the realm of the unvisualizable is the point at which macro-level or classical logic breaks down as well. As Max Jammer, the recognized authority on the history of quantum mechanics puts it, "It is hard to find in the history of physics two theories [wave and matrix mechanics] designed to cover the same range of experience which differ more radically than these two." Heisenberg characterized his view of the situation with the analogy that it is as if a box were "full and empty at the same time." Or, as Robert Oppenheimer put it, "If we ask, for instance, whether the position of the electron remains the same, we must say 'no'; if we ask whether the electron is at rest, we must say 'no'; if we ask whether it is in motion, we must say 'no.'"[7]

The confusion arises in part because of the classical assumption that all properties of a system, including those of microscopic atoms and molecules, are "real" in the sense that they are exactly definable and determinable. Bohr was among the first to realize that in the quantum world positions and momenta (where momentum is defined as the product of mass times velocity) cannot be said even in principle to have definite values. We deal rather in probabilities which in the Schrödinger formalism are expressed by the square of the amplitude of the wave function. It is only through acts of observation, which include us and our measuring instruments, that a definite value of the physical quantity is realized. This implies that the physical system does not exist in a well-defined state in which changes in the system occur continuously, and that the future state of this system cannot be predicted, even when the prediction is based on complete knowledge of the initial conditions.

Werner Heisenberg responded to the new situation with his famous indeterminacy principle. Put simply, this principle states that the product of the uncertainty in measuring the momentum, p, of a quantum particle times the uncertainty in measuring its position, x, is always greater than or equal to Planck's constant. In the language of mathematics, Heisenberg's principle is written as $\Delta x\, \Delta p \geq \hbar$, where the symbol Δx denotes the uncertainty in measuring the position x. Equally important, we confront here an indication that macro-level, or everyday, logic, which is premised on the law of the excluded middle, does not hold in the quantum domain. It is this realization, as we will see later on, that would lead Bohr to develop the new logical framework of complementarity.

At this point in the history of modern physics, physicists divided into two camps. Planck, Schrödinger, de Broglie, and, later, David Bohm joined ranks with Einstein in resisting the implications of quantum theory. Figures like Dirac, Pauli, Jordan, Born, and Heisenberg became, in contrast, advocates of the so-called Copenhagen Interpretation of quantum mechanics. Meanwhile, what would prove itself to be a complete theory, quantum mechanics, continued to be applied with remarkable success in its new form, quantum field theory. Here we witness the same correlation between increasingly elaborate mathematical descriptions of reality, a vision of the cosmos that is not visualizable, and the emergence of additional constructs that can only be understood within a new logical framework.

The New Logical Framework of Complementarity

This new logical framework, known as complementarity, will assume increasingly more importance in this discussion. Although it is a central feature of the Copenhagen Interpretation, which is now held to be the "orthodox" or standard interpretation of quantum theory, we will make the case that it is not well understood by many physical scientists, and also that it can be more generally applied than we have previously imagined. The unsettling conclusion forced upon us by complementarity, as Bohr and others have understood it, is that the truths of science are not "revealed" truths. We are now obliged to introduce into our understanding of the character of these truths an assumption that scientists have traditionally regarded as having no place or function in a scientific world view—the assumption is that scientific truths are, like other truths, subjectively-based constructs which are useful to the extent that they help us coordinate greater ranges of experience with physical reality. This last qualification means, as we shall see, a good deal, and allows us to avoid the conclusion that the subjective truths of science must be viewed as arbitrary in the sense that non-scientific truths are arbitrary. This is anything but the case. But this does not, as we shall also see, make this conclusion any less oppressive for many physical scientists.

Most physical scientists who have struggled with Bohr's alternate understanding of the character of scientific truths have appealed to the "wait and see" attitude described earlier. The tendency has been to relegate much of his commentary to a file drawer called "philosophy" in the hope that it would remain there until the foundations for his views in physical theory and experiment were obviated by further progress in both areas. Not only has the actual progress of physics failed to fulfill this hope, Bell's theorem and the experiments testing that theory seem to demand that we open that drawer and review its contents. As the physicist and philosopher of science Clifford Hooker notes, "Bohr's unique views are almost universally either overlooked completely or distorted beyond all recognition—this by philosophers of science and scientists alike."[8]

Our own attempt to better understand Bohr's views, with the assistance of the astute commentary of figures like Hooker, has led to some extraordinary conclusions about the heretofore largely unexamined role of complementarity in the history of modern physical theories, and the role that it could play in the future advance of physical theories in the field of cosmology, where one of us has particular expertise. We have also followed Bohr's suggestions about possible applications of the logical framework of complementarity to understanding fundamental logical relationships in knowledge fields outside of quantum physics by appealing to more recent knowledge that was not available to him. Some of the conclusions that we have drawn here, particularly in regard to Bohr's thesis about the role of complementarity in all conscious constructions of reality, must be viewed, at this stage in our quest for more complete scientific knowledge, merely as "hypotheses" which may be worthy of further investigation. As far as the major hypothesis in this discussion is concerned, however, there now appears to be an abundance of evidence in physical theory and experiments which appears to confirm it. This hypothesis concerns the manner in which complementarity serves to explain the new relationship between part and whole in modern physical theory in a quantum mechanical universe in which we must also now confront and deal with the amazing new fact of nature called non-locality.

The next chapter on the quantum mechanical view of nature, and the one that follows on Bell's theorem and the experiments testing that theorem, probably have more entertainment value in that we confront here more directly that which appears more unusual or bizarre from the point of view of everyday, visualizable reality. The entrance fee that the uninitiated must pay to journey into the brave new world of unvisualizable physical reality disclosed by quantum theory and experiment is an acceptance of the view that the truths of science have greater authority than the collection of arbitrarily developed assumptions known as "common sense." Since these scientific truths strongly suggest that we live in a quantum mechanical universe, the character of physical reality as we know it in quantum physics represents our best understanding to date of the conditions for our being and becoming in the vast cosmos. However odd or strange this reality might initially appear, it constitutes, from a scientific point of view, the "way things are."

2
The Strange New World of the Quantum

> The paradox is only a conflict between reality and your feeling of what reality ought to be.
> —*Richard Feynman*

Wave Particle Dualism and the Quantum Measurement Problem

The logical framework of complementarity was formulated originally in an effort to resolve some ambiguities concerning wave-particle dualism in quantum physics. Although new and profound complementary relationships, as we will discuss in more detail later, were also emergent in relativity theory, it was quantum physics which established that complementarity might be a fundamental aspect of our efforts to better comprehend physical reality with advances in physical theory. In classical physics, as we have seen, the visualizable physical reality appeared to be disclosed with ultimate completeness and certainty in physical theory. Thus physical theory, as Kelvin implied in his use of the "two cloud" metaphor, seemed in the actual practice of physics to bridge the gap between mind and nature and, in the process, to disclose the pre-existent scientific truths which lay beyond the veil of appearances.

The central problem in quantum physics is that the classical distinction between observer and observed system does not hold, and this results in the lack of a one-to-one correspondence between every element of the physical theory and the observed physical reality. What troubled physicists about this prospect was not that the reality disclosed by quantum physics could not be visualized based on experience in everyday visualizable reality—the unvisualizable character of physical reality in modern physical theory had already been convincingly demonstrated by relativity theory. What was troubling here was the suggestion that we can no longer "see" the pre-existent truths of physical reality through the lenses of physical theory in the classical sense.

The essential paradox of wave-particle dualism as visualizable phenomena is easily demonstrated. View the particle as a point-like something, like the period at the end of this sentence, and the wave as continuous and spread out. The obvious logical problem is how can a particular something localized in space and time, the particle, also be the spread out and continuous something, the wave? Quantum mechanics not only says unequivocally that quanta exhibit both properties, but also provides mathematical formalism governing what we can possibly observe, or see, when we coordinate our experience with this reality in actual experiments. In quantum physics, observational conditions and results are such that we cannot presume a categorical distinction between the observer and the observing apparatus, or between the mind of the physicist and the results of physical experi-

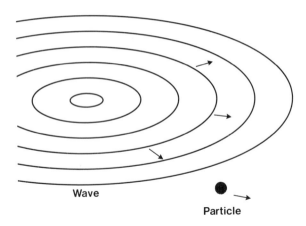

FIGURE 8. Wave and Particle. A wave is not localized whereas a particle is localized.

ments. The measuring apparatus and the existence of an observer are essential aspects of the act of observation in the quantum domain.

Understanding why this is the case, and why physicists were so dismayed that it should be the case, requires an understanding of the special character of the two theories, wave mechanics and matrix mechanics, which are now understood as complementary aspects of our description of this reality. The wave aspect of this reality is responsible for the formation of interference patterns. Interference, as we noted earlier, results when two waves, interacting with each other, produce peaks in places where they are combined, and troughs where they cancel each other. The height of the wave is termed its amplitude. One of the fundamental problems dealt with by quantum mechanics is to indicate where within the wave aspect of this reality we can hope to observe its particle aspect. Although we can predict the probability of finding a particle in a given region by squaring the amplitude of the wave function, $|\psi|$, *we cannot, even in principle, say precisely where the particle will be found prior to the act of observation.* This means that what is observed is fundamentally ambiguous unless we include the measuring apparatus and our-

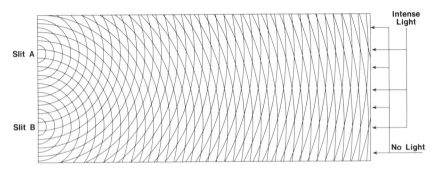

FIGURE 9. A reproduction of Thomas Young's original drawing showing interference effects of two waves originating in slits A and B.

FIGURE 10. The square of the wave function gives the probability of finding the particle somewhere prior to the act of measurement. After the measurement, the wave function is said to "collapse" and the particle is found at a specific location in space.

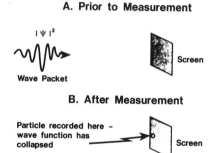

A. Prior to Measurement

$| \Psi |^2$

Wave Packet

Screen

B. After Measurement

Particle recorded here – wave function has collapsed

Screen

selves as part of the conditions of the experiment, or as primary in the analysis of the quantum results.

Although the square of the wave function gives the probability of finding the particle within a range of probabilities prior to the act of measurement, an actual measurement results in a situation in which the other "mathematically real" possibilities described by the wave function reduce to one set of possibilities. Since some possibilities are realized and others not, the wave function, in the language of physics, is said to "collapse" upon the act of observation, or when the particle is actually observed at a specific location in space.

In quantum physics, the total reality represented by the quantum system is dealt with in terms of two quite different theories—wave mechanics and matrix mechanics. As Bohr was among the first to realize, the relationship between these two theories and the reality which they describe makes sense only if we view them in terms of a new logical framework. Bohr began with the assumption that the wave aspect of this reality and the particle aspect are logically antithetical—something spread out and continuous in space-time represents a profound opposition to something with a discrete and well-defined location in space-time. Thus one view of the situation in a given instance must displace the other. Although the two theories are mathematically isomorphic, the choices that we are obliged to make in order to apply one theory will produce results that are logically disparate from those revealed in the application of the other. What makes the logical framework of complementarity new, or where it extends itself beyond our usual understanding of the character of logically antithetical constructs, is the following stipulation: In addition to representing profound oppositions which preclude one another in a given situation, both constructs are "necessary" for a complete understanding of the entire situation.

This is, as we will now try to demonstrate, pristinely true of wave mechanics and matrix mechanics. Wave mechanics is completely deterministic, and describes the continuous movement in time of a multidimensional spread out wave. It is complete in the classical sense in that it describes everything that can possibly be known about the quantum system in the "absence" of observation. If the quantum physicist confines himself to calculating the "mathematically real" possibilities given by the wave function, and is not required to demonstrate that all these "real" possibilities can be disclosed in a single experimental situation, wave mechanics appears to be the conceptual lens that allows us to "see" into the essences of this reality.

The initial impulse of Schrödinger, de Broglie, and others was to view the wave function as an actually existing entity like ordinary waves in water. The problem became, however, that although the wave function "theoretically" describes everything that can possibly happen in a quantum system, the actual observation of the system must deal in only the "probability" of finding a "something," or a quantum, at specific locations in space and in a specific energy state.

The other aspect of the complete quantum theory, matrix mechanics, deals, in contrast, only with the measurable quantities, or confines its seeing to the actual observables. A matrix is essentially a generalization of the idea of a simple number to a square or rectangular array of numbers, and the matrix that results when two matrices are multiplied depends on the order of the multiplication. For example, take a 2×2 matrix called A,

where $A = \begin{pmatrix} 1 & 0 \\ 2 & -3 \end{pmatrix}$

and multiply it by another 2×2 matrix called B,

where $B = \begin{pmatrix} 2 & -1 \\ 0 & 1 \end{pmatrix}$.

The matrix product $A \cdot B$ is another 2×2 matrix C, which can be computed using the rules of linear algebra

as $C = \begin{pmatrix} 2 & -1 \\ 4 & -5 \end{pmatrix}$.

The matrix product $B \cdot A$ is, however, a different 2×2 matrix D,

where $D = \begin{pmatrix} 0 & 3 \\ 2 & -3 \end{pmatrix}$.

Thus it is obvious that for these matrices that $A \cdot B \neq B \cdot A$. In the language of mathematics, matrices do not in general "commute." The formalism of matrix mechanics treats variables, such as position and momentum, as matrices. Since position and the corresponding momentum in a quantum system along the same direction do not commute, due to the presence of the quantum of action, matrices are useful tools for representing observables in quantum physics. What holds for matrices is translated in matrix mechanics into the formalism of linear algebra. The presence of non-commuting variables, such as position and momentum, means in this new formalism that we cannot know their values with a high degree of accuracy. If, then, we look at the quantum reality from the perspective of wave mechanics, we have a mathematical formalism describing a wave propagating deterministically with well-defined characteristics. In the absence of observation, there is a complete correspondence between every element of the physical theory and the physical reality in the classical sense. When, however, we make a measurement to determine what is actually there, some of the possibilities appear, and others disappear.

The confusion arises in part because one aspect of our description of this complete reality is classical. If a quantum system is left alone, meaning we do not attempt to observe it, the properties of the system change causally in accordance with the deterministic wave equation, like a system described in classical physics.[1] Yet another aspect of this reality, which is invoked when a measurement of the system is made, reveals that change in the system is discontinuous in accordance with the laws of probability theory. As physicist Eugene P. Wigner has emphasized, this is the most fundamental dualism encountered in quantum theory.[2] On the one hand, we have a classical system in which unrestricted causality and complete correspondence between every element in the physical theory and the physical reality holds. Yet, on the other, we have a completely non-classical system which features discontinuous processes, the absence of unrestricted causality, and the lack of a complete correspondence between physical theory and physical reality.

The choice of the phrase "collapse of the wave function" is unfortunate in that it implies that the wave function, as the term matter-wave initially suggested, is a real or actual something which exists unto itself prior to the act of observation, or in the absence of observation.[3] Viewing the wave function in this way requires that we assume that some aspects of this system, which were real or actual prior to, or in the absence of observation, somehow collapse or "disappear" when observation occurs. The quantum formalism in the Copenhagen Interpretation says nothing of the kind. What this formalism indicates is that prior to measurement we only have a range of possibilities, given by the wave function, in terms of mathematically derivable probabilities, given by the square of its amplitude, $|\psi|^2$. When an actual measurement is made, or when something "definite" is recorded by our instruments, the various possibilities become one "actuality." The wave equation of Schrödinger, which describes the evolution in space and time of the wave function in a totally deterministic fashion, cannot tell us what will actually occur when the system is observed. What the wave function provides is some indication of the range of possibility when a measurement takes place.

What has most troubled physicists is that one aspect of this reality as it is described in physical theory suggests that we have a complete theory which mirrors the behavior of the physical reality. Yet our efforts to coordinate experience with the total reality requires the use of another, logically disparate, theory as well. And there is no mechanism that allows us to predict precisely how or where the so-called collapse of the wave function will occur. The large problem, based on classical assumptions about the relationship between a "complete" physical theory and the physical reality described by this theory, is that an allegedly complete physical theory, quantum mechanics, does not and cannot allow us to describe when and how the collapse of the wave function occurs.

This enigma is reflected in two versions of quantum measurement theory. In the orthodox Copenhagen Interpretation, the wave function is viewed as a mathematical device, or idealization, which expresses the relationship between the quantum system, which is inaccessible to the observer, and the measuring device, which conforms to classical physics. What seems confusing here, particularly given the fact that we appear to live in a quantum universe, is the requirement that we view quantum reality with one set of assumptions, those of quantum physics, and the results of experiments in this physics as they are recorded on our measuring instruments with another set of assumptions, those of classical physics. This implies a categorical distinction between the micro and macro worlds, and yet does not specify at what point a measuring device ceases to be classical and becomes quantum mechanical. Does a system cease to be classical when it contains a billion, a trillion, or 10^{24} atoms? Add to this the obvious fact that any macroscopic system is made up of a multitude of particles obeying quantum physics, and the problem begins to seem even more irresolvable.

This "two domain" distinction in the orthodox quantum measurement theory has led, as we shall see, to enormous confusion about the character of quantum reality as well as to some conceptions of the character of this reality in physical theories that, initially at least, seem quite bizarre. The fundamental problem here, as we try to demonstrate in some detail later in our discussion of Bohr's views on quantum theory, is that the distinction between quantum and classical domains does not exist in nature. Bohr was among the first to fully accept the proposition that we live, in fact, in a quantum universe, and, therefore, that it is quantum physics which constitutes the most complete description of this universe. Yet Bohr's conclusion that we live in a quantum mechanical universe has not been widely recognized because he also suggested that our attempts to coordinate experience with this universe in physical theory requires the use of classical categories. His rationale for the use of those categories, which is rarely if ever mentioned in even advanced physics textbooks, is based on some profound revisions in traditional thinking about physics required by the observational conditions and results in quantum physics. And the implications of the new way of looking at the universe are, as we have suggested earlier, quite staggering. The reasons why Bohr's reading of the new situation in physics is appropriate, as well as why physicists have often chosen to resist acceptance of the new implications of quantum theory, will be explored in later chapters.

The second version of quantum measurement theory was proposed by John von Neumann in 1932 in an extended work entitled *Mathematische Grundlagen der Quantenmechanik*. In this version, the assumption is made that both the quantum system and the measurement devices are describable in terms of what Bohr viewed as only one complementary aspect of the total reality—the wave function. The decision to confer reality on the wave function in the classical sense, which has also been made in some actual physical theories with large cosmological implications, inevitably leads, as Bohr said it would, to unacceptable levels of ambiguity if not absurdity. In the absence in quantum physics of a mechanistic description of when and how the collapse of the wave function occurs, von Neumann concluded that it must occur in the consciousness of human beings. This refusal to view wave and particle as complementary aspects of the total reality, which experimental conditions and results in quantum physics clearly require, is, as we will try to demonstrate in some detail later, intimately connected with the impulse to preserve and protect the classical assumption that physical theory pristinely bridges the gap between mind and nature.

The Two Slit Experiments

One of the easiest ways to demonstrate that wave-particle dualism is a fundamental dynamic of the life of nature is to examine the results of the now famous two slit experiment. As physicist Richard Feynman put it, "any other situation in quantum mechanics, it turns out, can always be explained by saying, 'You remember the case with the experiment with two holes? It's the same thing.'"[4] Let us say that in our idealized experiment we have a source of quanta, in this case electrons, an electron gun, like that in TV sets, and a screen with two openings that are small enough to be comparable with the de Broglie wavelength of an electron. (See Plate I, p. xi.) Our detector is a second screen, like a TV screen, which flashes when an electron impacts on it. The apparatus allows us to record where and when an electron hits the detector.

With both slits S_2 and S_3 open, each becomes a source of waves which spread out spherically, come together, and produce interference patterns—bands of light and dark—on our detector. In terms of the wave picture, the dark stripes reveal where the waves have canceled each other out, and the light stripes where they have reinforced one another. If we close one of the openings, there is a bright spot on the detector in line with the other opening. The bright spot results from electrons impacting the screen in direct line with the electron gun and the opening. This result can be understood by viewing electrons as particles—we see no interference patterns or no wave aspect.

Physics has recently provided us the means of conducting this experiment with a single particle and its associated wave packet arriving one at a time. Viewing a single electron as a particle, or as a point-like something, we expect it, with both slits open, to go through one slit or the other—how could a single, defined something go through both? But if we conduct our experiment many times with both slits open, we see a buildup of the interference patterns associated with waves. Since the single particle has behaved like a wave with both slits open, it does, in fact, reveal its wave aspect. And we have no way of knowing which slit the supposedly particle-like electron passed through.

Suppose that we refine our experiment a bit more and attempt to determine which of the slits a particular electron passes through by putting a detector (D_2 and D_3) at each slit (S_2 and S_3). After we allow many electrons to pass through the slits, knowing which slit each electron has passed through, we discover two bright spots

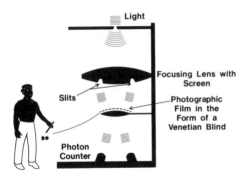

FIGURE 11. The delayed-choice version of the double-slit experiment of light according to Wheeler's thought experiment.

in direct line with each opening that the counter indicated the electron passed through. This is consistent with the particle aspect of the electron, and no interference patterns associated with the wave aspect are observed. Yet the choice to measure or observe what happens at the two slits reveals only the particle aspect of the total reality.

Now let us try to manipulate this reality into revealing one aspect or the other of itself by making extremely rapid changes in our experimental apparatus. The new experiment involves the double-slit arrangement with one modification—the photographic plate itself is sliced so that it acts as a venetian blind which is capable of creating an interference pattern when closed. When opened, it allows one photon at a time to go through and register a "click" on one of two photon counters located behind the venetian blind-like plate. This arrangement was originally proposed in 1978 by the physicist John A. Wheeler in a thought experiment known as the "delayed-choice" experiment. What we are doing here is opening or closing the venetian blind-like plate "after" the particle-like aspect has passed through the screen with the two slits.

According to the predictions of the thought experiment, when the blind is closed we should find that the photon has passed through both slits, and that the screen should register the wave aspect. But when the screen is opened, we should find that the photon makes only one click after interacting with one of the two photon detectors. In this case, the experiment predicts that we would conclude that the photon went through only one slit as we register the particle aspect. As Wheeler puts it "we decide, after the photon has passed through the screen, whether it shall have passed through the screen."[5]

Although the delayed-choice experiment was originally merely a thought experiment, we have been able to conduct actual experiments with single photons following two paths, or one path, according to a choice made "after" the photon has followed one or both paths. The experimental arrangement which actualizes the thought experiment, and confirms its predictions, allows the choosing to be done by an electronic, ultrafast Pockels cell.[6] Two groups, consisting of experimental physicists Carroll O. Alley, Oleg G. Jakubowicz, and William C. Wickers at the University of Maryland, and of T. Hellmuth, H. Walter, and Arthur G. Zajonc at the University of Münich, found that Wheeler's predictions were borne

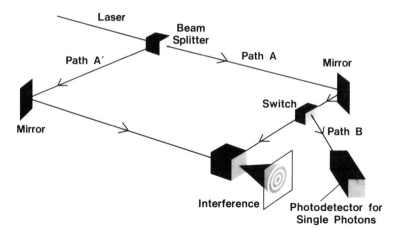

FIGURE 12. A delayed-choice experiment that has been carried out in the laboratory by groups at the University of Maryland and the University of München.

out in the laboratory. These results indicate that wave-like or particle-like properties are determined not just by the status of the two paths. They are also determined by the decision of the experimenter to make a measurement or observation by changing that status.

The results of these experiments not only clearly show that the observer and the observed system cannot be separate and distinct in space. They also show that this distinction does not exist in time. It is as if, no matter how strange it may sound, we "caused" something to happen "after" it has already occurred. These experiments, like those testing Bell's theorem which we will discuss in the next chapter, unambiguously disclose yet another of the "strange" aspects of the quantum world—the past is inexorably mixed with the present in a temporal non-locality. The neat classical notion of linear time, in which time present is inexorably followed by time future, breaks down. Even the phenomenon of time is tied to specific experimental choices.

For the non-physicist, it is not immediately obvious what an experiment with electrons, or photons, passing through two slits can possibly say about the vast complexity of the universe in which we live. The explanation is that what is disclosed in these experiments are "general" properties of all quanta, and thus fundamental aspects of "everything" in physical reality. We are not normally aware of quantum phenomena for the simple reason that they are quite small in comparison with our normative view of the nature of substance in everyday experience. For the physicist, trusting in the results of repeatable scientific experiments under controlled conditions, these properties are "real."

Planck's Constant

The central feature of the reality disclosed in the two slit experiments which allows us to account for the results is Planck's quantum of action. As Planck, Einstein, and Bohr showed, a change or transition on the micro level always occurs in terms of a specific chunk of energy. Nature is quite adamant about this, and there is no in-between amount of energy involved. Less than the specific chunk of energy means no transitions, and it is only whole chunks that are involved in quantum transitions. It is Planck's constant which weds the logically disparate constructs of wave and particle.

Let us illustrate this by returning to the double-slit experiment performed with a beam of electrons falling on a screen with the two openings, and suppose that we will now try to predict with the utmost accuracy the position and momentum of one electron. In quantum mechanics we find the momentum of a particle by taking Planck's constant, and dividing it by the wavelength λ of the wave packet representing that particle. A pure sinusoidal wave with a unique wavelength would, then, have a well-defined momentum. The problem is that such a wave would not be localized in any region of space and would, therefore, fill all space. Knowing the momentum precisely renders the position of the particle totally unknown. On the other hand, suppose we tried to isolate the quantum by confining it to a smaller and smaller wave packet. The problem with this strategy is that as we confine the wave aspect to increasingly smaller dimensions we have to include an increasing number of waves of different wavelengths added together. The mixture of wavelengths means that we are dealing with a mixture of momenta. In other words, as we make the wave packet smaller to correspond with the dimensions of the electron, we confront the general property of waves that in making the wave packet smaller we introduce more waves and, consequently, make momentum less precise.

This is where Planck's constant, or the rule that all quantum events occur in terms of this specific amount of action equal to 6.6×10^{-27} ergs sec, enters the picture. If Planck's constant were zero, there would be no indeterminacy because we could predict both momentum and position with the utmost accuracy. A particle would have no wave properties, and a wave no particle properties—mathematical map and the corresponding physical landscape would be in perfect accord. But Planck's constant is not zero. Mathematical analysis indicates that the "spread," or uncertainty, in position times the "spread," or uncertainty, of momentum is, greater than, or possibly equal to, the value of the constant or, more accurately, Planck's constant divided by 2π. If we choose to know momentum exactly, we don't know the position and vice versa.

Most important, the presence of Planck's constant means that we confront at the quantum level a situation in which the mathematical theory does not allow precise prediction of, or exist in exact correspondence with, the physical reality. If nature did not insist on making changes or transitions in these precise chunks, or in multiples of Planck's quantum of action, there would be no crisis—the universe, in principle, would be completely deterministic, and we could continue to assume that an exact one-to-one correspondence existed between every element of the

FIGURE 13. Illustration of Heisenberg's indeterminacy principle: (uncertainty in position) times (uncertainty in momentum) is at least as large as Planck's constant divided by 2π.

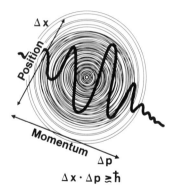

physical theory and the physical reality. Whether we view indeterminacy as a cancerous growth in the body of an otherwise perfect knowledge of the physical world, or the grounds for believing, in principle at least, in human freedom, one thing appears certain—it is an indelible feature of our understanding of nature.

In order to further demonstrate how fundamental the quantum of action is to our present understanding of the life of nature, let us attempt to do what quantum physics says we cannot do and visualize its role in the simplest of all atoms—the hydrogen atom. Imagine that you are standing at the center of the Houston Astrodome at roughly where the pitcher's mound is located. Place a grain of salt on the mound, and picture a speck of dust moving furiously around the outside of the dome in full circle around the grain of salt. This represents roughly the relative size of the nucleus and the distance between electron and nucleus inside the hydrogen atom when imaged in its particle aspect. It was, as we have seen, Rutherford's experiments which first suggested that matter, even apparently solid matter, is mostly empty.

In quantum physics, however, the hydrogen atom cannot be visualized with macro-level analogies like solar systems or the Houston Astrodome. In the original Bohr model, the electron was perceived as orbiting the positively charged nucleus at precisely specified orbits, with the space between orbits empty. Quantum theory as it developed in the late 1920s presented a different view—the vast space between the electron and the nucleus of the hydrogen atom is not "empty" in the classical sense. Each orbit in the Bohr atom is described by a probability distribution of finding the electron in an average position corresponding to each orbit. The space between these maxima of probability distributions is not empty—it is infused with energetic vibrations capable of manifesting themselves as quanta. The nucleus of the hydrogen atom and the electron can exchange quanta in the form of light, or photons, which are also produced by these energetic vibrations. It is wave mechanics which allows us to explain why electrons manifest at probabilistically determined energy levels at probabilistically determined distances from the nucleus. The energy levels manifest at certain distances because the transitions between orbits occur in terms of precise units involving Planck's constant.

This also means that if we attempt to observe or measure where the particle-like aspect of the electron is, as we did in the two slit experiment, the existence of Planck's constant will always prevent us from knowing precisely all the properties of that electron which we might presume to be there in the absence of measurement. And, as was also the case in the two slit experiment, our presence as observers and what we choose to measure or observe are inextricably linked to the results we get. Since all complex molecules are built up from simpler atoms, what we have said here about the hydrogen atom applies generally to all material substances. Any substance ultimately reduces to interactions of quanta in and between fields.

Quantum Probabilities and Statistics

What has most troubled physicists about quantum theory is, therefore, the implication that all the magnificent exactitude of mathematical physics has been reduced to calculating "probabilities" of events. A casual view of the situation would suggest that the grounds for objecting to quantum theory, the lack of a one-to-one correspondence between every element of the physical theory and the physical reality it describes, is justifiable and reasonable in strictly scientific terms. Since the "completeness" of all previous physical theories was measured against that criterion with enormous success, and since it was this success which gave physics the reputation of being able to disclose physical reality with magnificent exactitude, it may appear reasonable to assume that a more complete quantum theory might emerge by continuing to insist on this requirement. If we had been able, as Einstein fervently hoped, to circumvent quantum indeterminacy in such a way that the classical correspondence between physical theory and physical reality could still be assumed, there would be no reason to suspect that there might be some extra-scientific assumptions at work in this conception of scientific reality.

All indications are, however, that we cannot circumvent quantum indeterminacy in this way, and the success of quantum theory in coordinating our experience with nature is eloquent testimony to this conclusion. As Bohr was among the first to understand, the realization that we live in a quantum universe in which the quantum of action is a fundamental and irremedial fact requires a very different criterion for determining the completeness of physical theory. The new measure for a complete physical theory is that it must unambiguously confirm our ability to coordinate more experience with physical reality. If a theory does so and continues to do so, which is certainly the case with quantum physics, then the theory must be deemed complete. Science has always, of course, done its business in terms of this criterion, and this is precisely why we now confront a crisis if we insist on viewing the world in classical terms. Quantum physics not only works exceedingly well—it is, in these terms, the most accurate physical theory that has ever existed.

When we consider, for example, that this physics allows us to predict quantities like the magnetic moment of electrons to the fifteenth decimal place, and that these numbers have repeatedly been verified by experiment, we realize that accuracy per se is not the issue.[7] The issue, as Bohr rightly intuited, is that this complete physical theory effectively undermines the privileged relationship between mind and nature in physical theory which physicists have traditionally assumed to be both natural and necessary in the conduct of physics. It does so by indicating that there is no one-to-one correspondence between every element of the physical theory and the physical reality which allows one to "see" into the essences of this reality.

Another measure of success in physical theory, applied since the first scientific revolution, is also met by quantum physics—elegance and simplicity. The quantum recipe for computing the probabilities given by the wave function is straightforward and can be successfully employed by any undergraduate physics student—take the square of the wave amplitude, or the absolute value of the wave function squared, and thereby compute the probability of what can be measured or observed with a certain value. Yet there is, of course, a profound difference between the recipe for calculating quantum probabilities and the recipe for calculating probabilities in classical physics.

In quantum physics, for example, one calculates the probability of an event that can happen in alternative ways by adding the wave functions, and then taking the square of the amplitude.[8] In the two slit experiment, for example, the electron is described by one wave function if it goes through one slit, and by another wave function if it goes through the other slit. In order to compute the probability of where the electron is going to end up on the screen, we add the two wave functions, compute the absolute value of their sum, and square it. Although the recipe in classical probability theory appears as similar, it is quite different. In classical physics, one would just add the probabilities of the two alternate ways and let it go at that. The classical procedure does not work here because additional terms arise when the wave functions are added and the probability is computed in a process known as the *superposition principle*.

The superposition principle can be illustrated with an analogy from simple mathematics. In order to demonstrate how dissimilar the recipes are, consider the difference between adding two numbers and then taking the square of their sum, as opposed to just adding the squares of the two numbers. Obviously, $(2 + 3)^2$ is not equal to $2^2 + 3^2$. The former is 25, and the latter is 13. In the language of quantum probability theory

$$|\psi_1 + \psi_2|^2 \neq |\psi_1|^2 + |\psi_2|^2$$

where again ψ_1 and ψ_2 are the individual wave functions. On the left hand side one witnesses the superposition principle as extra terms arise which cannot be found on the right hand side. The left hand side of the above relation is the way a quantum physicist would compute probabilities, and the right hand is the classical analogue. In quantum theory, the right hand side is realized when we know, for example, which slit the electron went through.

Heisenberg was among the first to compute what would happen in an instance like this. The extra superposition terms contained in the left hand side of the above relation would not be there, and the peculiar wavelike interference pattern would disappear. The observed pattern on the final screen would, therefore, be what one would expect if electrons were behaving like bullets, and the final probability would just be the sum of the individual probabilities. Heisenberg provided a straightforward proof for this situation, which involves using the uncertainty relation.[9] When we "know" through which slit the electron went, this interaction with the system causes the interference pattern to disappear.

In order to give a full account of quantum recipes for computing probabilities, one has to examine what would happen in events that are compound. Compound events are "events that can be broken down into a series of steps, or events that consist of a number of things happening independently."[10] The recipe here calls for multiplying the individual wave functions, and then following the usual quantum recipe of taking the square of the amplitude, i.e., $|\psi_1 \cdot \psi_2|^2$. In this case, it would be exactly the same if we multiplied the individual probabilities, as one would in classical theory, for the simple reason that

$$|\psi_1 \cdot \psi_2| = |\psi_1| \cdot |\psi_2|.$$

Thus, the recipes of computing results in quantum theory and classical physics can be totally different. The quantum superposition effects are completely non-classical, and there is no mathematical justification per se why the quantum recipes work. What justifies the use of quantum probability theory is the same thing that justifies the use of quantum physics—it has allowed us in countless experiments to vastly extend our ability to coordinate experience with nature. The bottom line is that both work, and work beautifully.

Although it was not known at the time, the view of probability in classical probability theory was greatly conditioned by what now appears to be extra-scientific assumptions about the relationship between physical theory and physical reality. In the nineteenth century physicists developed sophisticated statistics to deal with large ensembles of particles before the actual character of these particles was understood. Classical statistics, developed primarily by James C. Maxwell and Ludwig Boltzmann, was used successfully to account for the behavior of molecules in a gas, and to predict the average speed of a gas molecule in terms of the temperature of the gas. Since nothing was known about quantum systems, and since quantum indeterminacy is small when dealing with macro-level effects, the presumption was that the statistical averages were workable approximations which subsequent physical theories, or better experimental techniques, would disclose with exact precision and certainty. This was, as we saw earlier, the manner in which Einstein was defining probability in his paper on Brownian motion. We now know that quantum mechanical effects are present in the behavior of gases, and that the choice to ignore them is merely a matter of convenience in getting workable or practical results. It is no longer possible to assume, in other words, that the statistical averages are merely higher-level approximations for a more exact description which would reveal a one-to-one correspondence between every element in a physical theory describing the behavior of gases and this physical reality. Moreover, what kind of statistics an ensemble of quanta or particles follows depends on the individual spins of the particles or quanta. Integer-spin particles can be crowded together in a single state but half-integer-spin particles cannot—the latter obey the Pauli exclusion principle. In either case, the resultant statistics are radically different from the classical statistics of Maxwell and Boltzmann.

The Schrödinger Cat Paradox

The first major defense of the classical conception of the relationship between physical theory and physical reality took the form of a thought experiment involving a cat. And this cat, like the fabulous beast invented by Lewis Carroll, appears to have become quite famous. The thought experiment, proposed by Schrödinger in 1935, is designed to parody some perceived limitations in quantum physics and, like many parodies in literature, the underlying intent was quite serious. In what had already become the orthodox interpretation of quantum theory, the Copenhagen Interpretation, the act of measurement plays a central role. Prior to the act of measurement, one cannot know which of the many possibilities implied by the wave function will be materialized. Schrödinger, the father of wave mechanics, was a believer, along with Einstein, in the one-to-one correspondence between every element of the physical theory and the physical reality. The intent of the thought experiment was to argue indirectly that mathematically real properties are "real" even in the absence of observation.

In this Rube Goldberg–like thought experiment, we must first imagine that Schrödinger's cat is a collection or ensemble of wave functions which correspond with the individual quantum particles. In other words, we must identify the "reality" of the cat with the wave function. The cat is first placed inside a sealed box that can release poisonous gas. The release of the gas is determined by the radioactive decay of an atom, or by the passage of a photon through a half-silvered mirror. Schrödinger chose to have the gas released in this way because either trigger is quantum mechanical and, therefore, indeterminate or random. The parody of the role of observation in orthodox quantum measurement theory takes the form of a question. Since the observer standing outside the box does not know when the gas is released, or if the cat is alive or dead, the question is, "What is happening inside the box in the absence of observation?" Although the intent was humorous, the principle at issue for Schrödinger and Einstein was very serious indeed. The thought experiment suggests that the cat must be both alive and dead prior to the act of observation since both possibilities remain in the isolated system in the absence of observation. Thus Schrödinger is indirectly suggesting, in the effort to point up the absurdity of any alternate view, that a mathematically real property exists in the physical reality whether we observe it or not.

The essential paradox Schrödinger seeks to amplify here has been nicely described by Abner Shimony:

> There would be nothing paradoxical in this state of affairs if the passage of the photon through the mirror were objectively definite but merely unknown prior to observation. The passage of the photon is, however, objectively indefinite. Hence the breaking of

**Cat is Alive –
50% Probability**

FIGURE 14. Schrödinger's cat in box thought experiment: There is a 50% probability at any time that the cat is dead or alive.

**Cat is Dead –
50% Probability**

the bottle is objectively indefinite, and so is the aliveness of the cat. In other words, the cat is suspended between life and death until it is observed.[11]

One might be able to dismiss the paradoxical nature of this conclusion if it were supported merely by a thought experiment. Yet here, as in the delayed-choice thought experiment of Wheeler, physicists have developed actual experiments to test the paradox. Groups at the IBM Thomas J. Watson Research Center, the AT&T Bell Laboratories, the University of California at Berkeley, and the State University of New York at Stony Brook have carried out experiments which are attempts to confirm Schrödinger's cat paradox. These experiments are based on calculations done by Anthony J. Leggett and Sudip Chakravarty, and involve the quantum "tunneling" effect. Quantum tunneling involves the penetration of an energy barrier, and is completely forbidden in classical physics. It accounts, among other things, for the radioactive decay of nuclei and for nuclear reactions. Quantum tunneling in these experiments takes place only if a physical quantity, the magnetic field in a superconducting ring, is indefinite, or in suspended animation. In analogy to the cat being both dead and alive, the magnetic flux does not have one or the other of the two possible values. It is important to realize here the magnetic field in this experiment is, like the cat, a macroscopic quantity. It is this which makes the analogy of the superconducting ring with Schrödinger's cat valid, and allows us to draw experimentally valid conclusions about the role of the observer as it is viewed in orthodox quantum measurement theory.

In the superconducting ring experiments the orthodox view holds—the state of the macroscopic system, namely, the conducting ring, cannot be determined until the measurement takes place. Moreover, one cannot even insist that the macroscopic state somehow has a definite value prior to the act of the observation. The state of the system is indeed "dependent" upon the act of observation. Its otherwise mathematically real possibilities, as given by the Schrödinger wave equation, "collapse" upon the act of observation. Yet the paradox, as we will try to show later, has nothing to do with alive or dead cats, or the equivalent state in superconducting rings. This distorted view arises out of a fundamental confusion about the character of reality-in-itself fostered by the classical assumption that a real or "objective" description of physical reality must show a one-to-one correspondence between the physical theory and the physical reality.

Future experiments of this kind will attempt to show whether the calculation of quantum probabilities for many events in a statistical sense are, like those in experiments testing Bell's theorem, dependent upon acts of observation. If the answer is yes, as we expect it will be, the experiments would indicate that the magnetic field would be in a superimposed state[12] prior to the act of observation, or that the field in analogy to Schrödinger's cat would be in a state of suspended animation. If we assume, as Schrödinger did, a one-to-one correspondence between every element of the physical theory and the physical reality, then we would face here the paradox of the cat that is somehow both alive and dead simultaneously. If we do not make that assumption, in accordance with Bohr's understanding of the actual conditions and results of quantum mechanical experiments, there is, as we shall see later, no such paradox.

The Schrödinger cat paradox and Wheeler's delayed-choice thought experiment and the laboratory experiments testing them indicate that, in contrast to a common misunderstanding of quantum theory, quantum indefiniteness or quantum uncertainties are not confined to the microscopic realm: They spill over to the macroscopic realm. This, we believe, has important consequences for understanding the character of physical reality at all scales as we will attempt to demonstrate later in our discussion of cosmological observations.

Quantum Field Theory

Contemporary physics is built on quantum mechanics which has been extended and refined into quantum field theory. Further developments in quantum theory led to the publication of a relativistic quantum theory in 1928 when Paul A.M. Dirac combined special relativity with quantum mechanics. The result was a theory that predicted the existence of positively charged electrons termed the positrons, or the anti-particles of regular electrons. The jewel of modern quantum field theory, quantum electrodynamics, or QED for short, was developed much later. It accounts for interactions of not just electrons and positrons, but also of other charged particles. Electromagnetic interactions are mediated by photons, and QED is, therefore, a quantum field theory of electromagnetic interactions. QED was fully developed in the 1940s, and one of its principal architects and proponents was Richard Feynman. Since the success of QED in dealing with electromagnetic interactions has been so great, the consensus is that it should provide a successful model for the less well understood strong interactions as well.

The concepts of fields and their associated quanta are now fundamental to our understanding of the character of physical reality. Yet these concepts, like that of four-dimensional space-time in relativity theory, are totally alien to everyday visualizable experience. Again, let us attempt what quantum physics deems impossible and try to visualize this unvisualizable reality. First imagine that the universe runs like a 3-D movie. What we can detect or measure in this movie are quanta, or particle-like entities. These quanta are associated with infinitely small vibrations in what can be pictured as a grid-lattice filling three-dimensional space. Potential vibrations at any point in a field are capable of producing quanta which can move about in space and interact. And increasingly higher and higher energies are present in smaller and smaller regions in space. It is the exchange of these quanta, the carriers of the field interactions, that allows the cosmic 3-D movie to emerge and evolve in time. There are four known field interactions—strong, electromagnetic, weak, and gravitational.

In quantum field theory particles are not acted upon, as classical physics supposed, by "forces"—they rather "interact" with each other through the exchange of other particles. The laboratories that have provided experimental evidence confirming and refining the predictions of quantum field theory are high-energy particle accelerators. Such devices have been described as the modern equivalent of cathedrals built in the twelfth and thirteenth centuries and with good reason. They are also costly and magnificent artifacts of belief in the wonder and extent of the universe. The main feature of these accelerators is a large hollow ring within which electrons or protons are accelerated to great speeds, and made to interact with other particles. The accelerators are not atom smashers in the sense of breaking up matter into smaller or more basic components. The effect of the collisions is rather "transformations" in which enormous energy briefly bursts open the world of fields, providing a backward look into the high energy regime that dominated the early life of the cosmos.

As we engineer higher energy in the accelerators simulating conditions in an earlier, much hotter universe, something remarkable happens—the fields begin to blend or to transform into more unified fields. The hypothesis is that given enough energy, which we cannot hope to produce in today's particle accelerators, we would be able to disclose conditions close to the point of origins of the cosmos where all the fields were unified into one fundamental field. Even if we do elect to construct in Texas the proposed superconducting supercollider with its fifty-three mile–long tunnel at a cost of $4.4 billion, the energies produced will not be sufficient to simulate conditions in the unified field. The general rule in physics which applies here is that increase in energy correlates with increase in symmetry, or in new patterns of interactions disclosing fewer contrasting elements. The expectation is that the ultimate symmetry in the cosmos at origins would reveal no contrasts or differences in an unimaginable oneness in which no thing, or nothing, exists to be observed or measured. It would be equivalent to what mathematicians call an empty set.

We used the analogy of the 3-D movie partially because our normative seeing is in three dimensions, as opposed to the four-dimensional reality of space-time that the theory of relativity and quantum field theory presume. But the primary motive for metaphor is that 3-D movies require that we put on glasses to view them. The putting on of glasses to observe the cosmic movie can be likened to acts of making observations or measurements of micro-level events. And the action in the movie that we might presume to be there in the absence of measurement, or before putting on the glasses, is not the same as that which we actually observe in physical experiments. Our problem in this movie house is that we are confronted with two

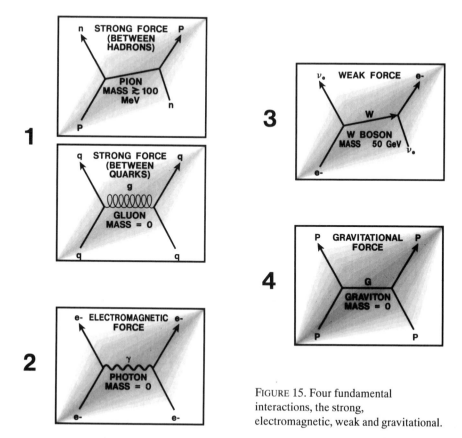

FIGURE 15. Four fundamental interactions, the strong, electromagnetic, weak and gravitational.

logically antithetical aspects of one complete drama, and the price of admission is that we cannot perceive or measure both simultaneously.

The central feature of quantum field theory, as Steven Weinberg has defined it, is that "the essential reality is a set of fields subject to the rules of special relativity and quantum mechanics; all else is derived as a consequence of the quantum dynamics of those fields."[13] Quantum field theory has also revealed an important complementarity between particles which are localizable in space-time and fields which are not. As the philosopher of science Errol E. Harris observes:

> Quantum theory coupled atomicity with radiation and linked momentum with position in inseparable complementarity. Particle physics today fits the particles it has distinguished inseparable from the fields of force they exert.[14]

Material reality as we have come to know it in quantum field theory is constituted by the transformation and organization of fields and their associated quanta. It is this which is the basis for the emergence and evolution of all events in the vast cosmos. The quantization of fields in the effort to comprehend this fundamental process is essentially an exercise in which we use complex mathematical models

Four Elementary Interactions Found in Nature

Interaction	Physical Phenomena	Range[a]	Relative Strength	Radiation[b] Quanta	Matter[c] Quanta
Strong	nuclear forces	10^{-13} cm	1	gluons	quarks[d]
Electromagnetic	atomic forces, optics, electricity	∞	10^{-2}	photon	quarks, leptons (i.e., electrons, muons)
Weak	radioactivity, nuclear reactions in stars	10^{-16} cm	10^{-10}	$W^{\pm}$, Z	quarks, leptons, neutrinos
Gravitational	planetary orbits, binary stars, galaxies, clusters of galaxies, black holes, etc.	∞	10^{-38}	graviton	everything

[a] 10^{-13} cm is about the size of a nucleus.
[b] These are the quanta which are responsible for the transfer of the interaction.
[c] These are the particles which interact with each other under the particular interaction.
[d] Quarks are the constituent particles of neutrons and protons and other heavy particles.

to analyze the field in terms of its associated quanta. If quantum field theory is as "rock bottom" or fundamental to the description of the actual dynamics of material reality as we now presume it to be, there is obviously little hope of ever describing "reality-in-itself" in physical theory. What appears as a continuous description of interactions of fields is interactions of quanta exchanging the carriers of the field. In the case of the interaction of electrons, we are dealing with the exchange of photons, the carriers of the electromagnetic field. At the nuclear level, particles like the proton and the neutron are composed of quarks linked together by the exchange of gluons, the carriers of the strong force between quarks. Quarks, in turn, make up all the heavy particles, the hadrons, such as the proton and the neutron. The fundamental reality of all these fields is described as potential vibrations at each point, which are capable of manifesting themselves in their complementary aspect as individual particles which mediate interaction between the continuum of fields.

What this picture suggests is that our previous distinction between the four fundamental interactions, gravity, electromagnetism, the weak, and strong forces, was superficial. Each of these interactions is mediated by quanta—the graviton in the case of gravity, the photon for electromagnetism, intermediate bosons for the weak force, and the "colored" gluons for the strong binding force. It is the mathematical description of these quanta that is provided by relativistic quantum field theory. What that description clearly indicates is that the four interactions become increasingly unified at ultrahigh energies, corresponding with those present in the very early life of the universe. If we could re-create the energies present in the first trillionths of trillionths of a second in the life of the universe, these four fields would, according to quantum field theory, become one fundamental field.

As we will suggest in more detail later, quantum field theory, like the rest of modern physics, discloses a profound new relationship between part and whole which is also completely non-classical. Physicists, in general, have not welcomed this new relationship primarily because it unambiguously suggests that the classical conception of the ability of physical theory to disclose the whole as a sum of its parts, or to "see" reality-in-itself, can no longer be held as valid. We are now forced to realize that physics is no longer the business of disclosing pre-existent truths, or, if you like, that such truths are not, as we once supposed, revealed truths. There is no choice, in our view, but to view the truths of science as subjective, and useful to the extent that they help us coordinate greater ranges of experience with physical reality. Yet the loss of this old conception of scientific truths has its compensations in the form of the alternate framework of complementarity. The compensation, as we will try to illustrate in the second half of this discussion, is more than adequate to cover this loss. Although the new epistemology of science indicates that we cannot, even in principle, know reality-in-itself in the old terms, it also provides a foundation for conceiving of our relationship to this reality which makes the business of being conscious human beings a good deal more meaningful than classical physics ever allowed.

As we suggested earlier, the effect of Bell's theorem and of the experiments testing that theorem has been to force resolution of the fundamental questions posed by the conditions and results of quantum mechanical experiments by framing those questions within the context of the actual practice of science. The best and perhaps easiest way to understand the questions Bell posed in his theorem is probably to revisit the central questions that were left unresolved in the twenty-three year–long debate between Einstein and Bohr. When Bell asked these questions in the form of a theorem which could serve as the basis for experiments which would provide answers, he did not know what the answers would be. The result that has most perplexed physicists was the discovery of a new fact of nature known as non-locality which, like virtually all of the facts associated with quantum physics, completely confounds assumptions about the character of the real world based on macro-level visualizable experience.

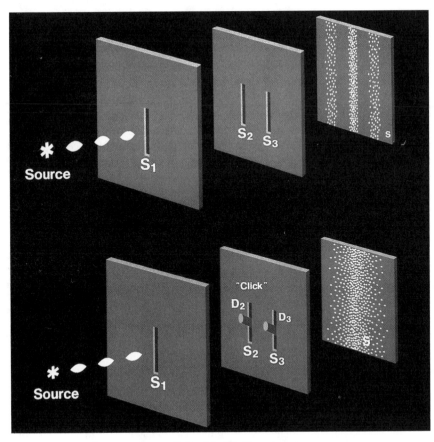

PLATE I. Double-Slit Experiment.

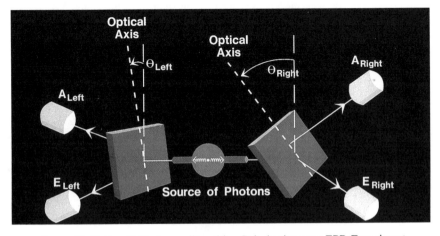

PLATE II. Testing Bell's Inequality with a Polarization-type EPR Experiment.

3
Bell's Theorem and the Aspect Experiments: Coming to Terms with Quantum Non-local Reality

The great debate over the world-view implied by quantum theory between Bohr and Einstein began at the fifth Solvay Congress in 1927, and continued intermittently until Einstein's death in 1955. The argument took the form of thought experiments in which Einstein would try to demonstrate that it was theoretically possible to measure, or at least determine precise values for, two complementary constructs in quantum physics, like position and momentum, simultaneously. Bohr would then respond with a careful analysis of the conditions and results in Einstein's theoretical examples, and demonstrate that there were fundamental ambiguities which the thought experiments failed to resolve. Although both men would have despised the use of the term, Bohr was the "winner" on all counts. Eventually, the dialogue revolved around the issue of "realism," and it is this issue that Einstein felt would decide the correctness of quantum theory. Although the second major phase of this debate began in 1935, the issue of realism was resolved in Bohr's favor only after the publication of the results of the Aspect experiments in 1982.

One of the early thought experiments proposed by Einstein, the so-called *clock in the box* experiment, illustrates how each stage of the debate typically played itself out. Suppose, Einstein suggested, we have a box that has a hole in one wall, and that this hole is covered by a shutter which can be opened and closed by the action of a clock inside the box. Assume that the box contains, in addition to the shutter mechanism and the clock, radiation, or photons of light. We now set up the apparatus so that the clock will open the shutter at some precise time, and thereby allow one photon, or quantum of light, to escape before it closes. In this thought experiment we weigh the box before the photon is released, wait for the photon to escape at the precise, predetermined time, and then weigh it again. Since mass is equivalent to energy, argued Einstein, the difference in the two weights will tell us the energy of the photon that escaped. This procedure should allow us to know, concluded Einstein, the exact energy of the photon and the exact time it escaped. Thus both of these allegedly complementary aspects of the system would be known, and the uncertainty principle refuted.

Bohr, focused as always on the conditions and results of experiments, showed why this procedure cannot produce the predicted result. Since the box must be weighted, it must be suspended by a spring in the gravitational field of the earth. As Einstein himself had demonstrated in his general theory of relativity, the rate at which the clock runs is dependent upon its position in the gravitational field. Bohr then pointed out that as the photon escapes, the change in weight and the recoil

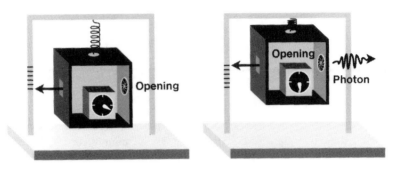

FIGURE 16. Clock in the box experiment. Thought experiment devised by Einstein to refute the uncertainty principle.

from the escaping photon would cause the spring to contract and, therefore, alter the position of both box and clock. Since the position of both changes, there is some uncertainty regarding position in the gravitational field, and, therefore, some uncertainty in the rate at which the clock runs. Suppose, however, we attempt to restore the original situation by adding a small weight to the box which would stretch the spring back to its original position, and then measure the extra weight to determine the energy of the escaping proton. This strategy will not work, said Bohr, because we cannot reduce the uncertainty beyond the limits allowed by the uncertainty principle, in this case

$$\Delta E \; \Delta t \geq \hbar.$$

The details of this and other thought experiments in the Bohr-Einstein debate can be found in *Subtle is the Lord* by Abraham Pais.[1] What is remarkable here is that although Einstein was enormously persistent in his efforts to disprove the uncertainty principle and, therefore, the Copenhagen Interpretation, he was also quite willing to accept the inadequacy of one thought experiment after another based on Bohr's detailed replies. What both tendencies illustrate is that Einstein knew full well that he was confronting dilemmas which dwarf any narrow concerns about professional reputation, or even the merits of a physical theory.

The EPR Thought Experiment

Eventually Einstein was willing to accept the uncertainty or indeterminacy principle. Although the essential point of subsequent disagreement in the debate was that Einstein, in contrast with Bohr, felt that quantum theory was not a complete theory, the more substantive point of disagreement involved some profound differences concerning the actual character of scientific epistemology, and, therefore, of the knowledge we call physics. The differences in background and experience which seemed to have allowed Bohr to be more prescient in his reading of the world-view of quantum physics will concern us briefly later in this discussion. At the moment we are interested only in the more narrowly scientific aspects of the dialogue in the thought experiment proposed by Einstein, Podolsky, and Rosen in 1935—the so-called EPR experiment.[2] It is this experiment which would lead to Bell's theorem and the experiments testing that theorem.

The EPR thought experiment was conceived in the early 1930s, and involves a new kind of imaginary test for orthodox quantum measurement theory. The impulse here was basically to use experimental information about one particle to deduce allegedly complementary properties, like position and momentum, of another particle. While at Princeton during 1934 and 1935, Einstein refined his argument in dialogue with Boris Podolsky and Nathan Rosen, and presented it in a paper published in 1935. The rationale for this thought experiment was the same as that in all the previous thought experiments devised by Einstein in the endless debate with Bohr. Quantum mechanics is "incomplete," alleged Einstein, Podolsky, and Rosen in the EPR paper, because it does not meet the following requirement— "every element in the physical theory must have a counterpart in the physical reality." What the physicists were seeking to protect and preserve was the one-to-one correspondence between the physical theory and the physical reality. It is belief in this correspondence, as we have seen, that allowed physicists to conclude that physical theory was the means through which one could "see" into the essences of physical reality.

In this thought experiment, we are asked to imagine that two particles originate from a definite quantum state, and then move apart without interaction with anything else until we elect to measure or observe one of them. Any quantum state is characterized by a set of quantum numbers, and two particles originating from this state display variables which obey the quantum rules for the quantum state under consideration. Electrons are, for example, characterized by quantum numbers which give the value of orbital angular momentum and their combined spin. If, for example, the initial state of an electron had zero orbital angular momentum and zero spin, the two particles would have equal and opposite spins to each other so that the total value of their spins would always be zero. In other words, *once a direction in space has been chosen, as for example when the spin of one particle is measured along a given axis in space*, the spins will show strict correlation. The spin of one particle originating in a single quantum state will be along the given direction in space, and the spin of the other particle will be exactly opposite to it. Such spin states in the language of physicists are called "singlet."

Since the quantum rules allow us, when the two particles are initially in a definite quantum state, to calculate their initial momentum, the EPR argument was that the individual momenta will be correlated even after the particles separate. We are asked in this thought experiment to measure the momentum of one particle after it has moved a significant distance from the other, or achieved what physicists call a "space-like separation" where no signal traveling at the upper limit of the speed of light can carry information between the separated particles. Presuming that the total momentum is conserved, we should be able, argued Einstein and his colleagues, to calculate the momentum of the other particle which we have not measured. Since our measurement of the momentum of the one particle invokes the quantum measurement problem, Einstein fully conceded that we cannot know the precise position of this same particle.

In spite of this limitation, Einstein assumed that measurement of the momentum of the particle we actually measured would not disturb the momentum of the other, which is space-like separated and which could also be as far away from the first as one likes. Since we know the momentum of the first of the two particles, the position measurement of the second coupled with the knowledge of the first particle's momentum should, claimed Einstein and his colleagues, allow us to deduce both momentum and position of the second particle. This means that we should be able to deduce both position and momentum for a single particle, which would violate the indeterminacy principle, and arrive at a one-to-one correspondence between physical theory and physical reality. The paper concludes that the orthodox Copenhagen Interpretation "makes the reality of [position and momentum in the second system] depend upon the process of measurement carried out on the first system which does not disturb the second system in any way. No *reasonable* definition of *reality* could be expected to permit this."[3] [emphasis ours]

Bohr countered that a measurement by proxy does not count, and that you cannot attribute both position and momentum to a single particle unless you measure that particle. But what would prove most important about the EPR thought experiment is that it featured another fundamental classical assumption which physicists at the time regarded as incontrovertible truth, and which actual EPR-like experiments have shown to be false. It is known as the principle of local causes. What the principle states is that a physical event cannot simultaneously influence another event without direct mediation such as the sending of a signal. This means in the EPR experiment that a measurement of one particle cannot simultaneously affect the measurement of the second particle in a space-like separated region. One would have to assume, if the principle of local causes is valid, that a signal can travel faster than light for such an influence to occur, and, consequently, abandon the theory of relativity and all of modern physics.

Bell's Theorem

What was needed to finally settle the central issue of realism in the Bohr-Einstein debate was a variant of the EPR experiment that would provide a direct test of whether the indeterminacy principle would be violated. John Bell of the Centre for European Nuclear Research conceived of a way to accomplish this in 1964. Beginning with the fundamental assumptions of the EPR experiment, one-to-one correspondence between the physical theory and the physical reality and the absence of faster-than-light communication, Bell deduced, mathematically, the

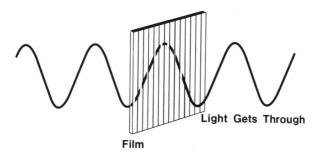

Light Gets Through

Film

FIGURE 17. Illustration of the polarization of light measured with a piece of polarized film. Light gets through if it is polarized along the transmission axis of the film.

most general relationships between two particles like those in the EPR experiment. What he proved is that certain kinds of measurement could distinguish between the positions of Einstein and Bohr. One set of experimental results would prove quantum theory "complete" and Bohr correct, and another set would prove quantum theory incomplete and Einstein correct.

Although modern versions of the EPR experiment involve measurements of properties of light, like polarization, as opposed to properties of particles with mass, like position and momentum, the fundamental issues involved are precisely the same. In the absence of experimental results in the experiments suggested by his theorem, Bell himself had no idea of what the conclusion would be. The mathematical statement derived in Bell's theorem is known as Bell's inequality. Local realistic theories would obey the inequality and, therefore, if the inequality held in experiment, Einstein would be correct. If violated, quantum theoretical predictions would be valid and Bohr would be the victor. The important point is that the issue could now be submitted to the court of last resort—repeatable scientific experiments under controlled conditions.

In order to understand the character of Bell's theorem, it is first necessary to be clear about what spin and polarization mean. First of all, many of the peculiarities of the spin of a particle, like an electron, can be ignored in modern EPR-like experiments. If, for example, a particle "turns around" before it shows the same face again, this does not matter because the spin of the particle defines a certain direction in space, not unlike the manner in which the spin of the earth defines the direction of the north-south axis. When brought into a uniform magnetic field, the electron will line up in one of two possible states. It will be parallel to the field or anti-parallel, or, in the terms normally used to describe this orientation, either "up" or "down." Assuming that the electrons originate in a single quantum state of zero angular momentum, the total angular momentum of this pair of photons would be zero. Although each of the electrons can have angular momentum, or spin, they must have equal and opposites amounts of spin in order that the total for the pair remains zero. This is always the sum required if the two electrons were formed initially in a single quantum system of zero angular momentum.

The quantum dilemma arises here when, for example, we try to determine the spin of a single particle in this pair. In classical terms, the assumption is that we are dealing with particles in three-dimensional space, and must, therefore, measure three directions of spin. Adding the three components together should, therefore, allow us to arrive at the total spin. The quantum reality does not, however, allow this simple procedure to work. If we measure one component of spin, the others can only be determined to the extent allowed by the uncertainty principle. As with position and momentum, the spin components in two directions are related to each other through the uncertainty principle—the spin vectors are complementary pairs. They cannot be measured simultaneously any more than position and momentum can be measured simultaneously. Since the spin of a particle like an electron is quantized, a measurement of the spin in any direction will only yield the result "up" or "down," often written as +1 or −1. If we measure the spin in a given direction on one axis, we have a 50% chance that we will get the value +1 and equal probability for the value −1. If we elect to measure the spin in the other direction along the second axis, we are now forced to take the results of the first measurement into account. Even if we knew that the measurement on the first axis was, say, "up" before we made a measurement in the second axis, the measurement taken in the other direction would yield the answer "up" over repeated measurements only 50% of the time. In other words, the complementary character of this reality is such that quantum uncertainty is adhered to each time we make a measurement.

When we view the particles in isolation, each particle appears to be undergoing random fluctuations which would confuse attempts to know in advance the spin of a single particle. Yet we also know that the two particles in the paired system have equal and opposite spin. This means that the random spins in one particle will match *precisely*, or correlate with, those of the other. What troubled Einstein here is that the quantum formalism indicates that this correlation should be present regardless of the distance between the two particles. The formalism predicts the correlations will exist even if the distance between the particles is such that a faster-than-light signal would seem necessary to account for the correlations. This was an example of what he termed "spooky action at a distance."[4]

Although most of the experiments testing Bell's theorem involve polarization of photons, rather than spin of electrons, the principles at work, and at issue, are precisely the same. Polarization defines a direction in space associated with the wave-aspect of a photon just as spin defines a direction in space for a material particle like the electron. The double-slit experiments, which we used earlier to illustrate the peculiar properties of the quantum world, feature polarization, and these same peculiar properties are in evidence in the experiments testing Bell's theorem. Although we were concerned earlier only with the examples of horizontal and vertical polarization, there is also a kind of polarization known as circular polarization. This form of polarization can be crudely visualized as a twirling baton which has a particular orientation as it moves. Circular polarization is either left-handed or right-handed, and has been particularly useful in experiments testing Bell's theorem.

The polarization of a photon, like the spin of an electron, has, therefore, a "yes" or "no" property, or will manifest in photons which are part of the same quantum system in complementary fashion. With the complementary nature of polarization in mind, we can illustrate the results of the experiments testing Bell's theorem with a simple two-photon system which uses a crystal similar to a polarizing film as a transmission device.[5] Such a crystal splits a beam of light that falls on it into a beam which is linearly polarized along the axis of the crystal and another beam polarized perpendicular to the axis of the crystal. Detectors record the path of each photon correlating with either parallel or perpendicular polarization.

Since quantum theory predicts the probabilities of each possible experimental outcome when the photon is polarized along the optical axis of the crystal, the probability that it will pass through the crystal and be recorded along that channel, for the reasons we have seen, is 1. If it is polarized perpendicular to the optical axis of the crystal, the probability for passing through it and being recorded along the same channel is zero. Quantum theory also predicts that if the photon is linearly polarized at some angle θ to the transmission axis, where θ is between 0 and 90 degrees, that the probability of photons passing through the crystal is a number between 1 and 0. Specifically, quantum theory predicts that the probability is equal to the square of the cosine of the angle, or $(\cos\theta)^2$.

Now suppose, as in the original EPR argument, that two photons originate from a single quantum state, and propagate in two opposite directions. In certain quantum states, each beam by itself appears completely unpolarized. However, each photon's polarization is perfectly correlated with its partner's. In other words, the total polarization of the two-photon system is such that the two individual polarizations would always have to be along the same direction in space. Thus one possible quantum state of the pair of photons is the state in which both photons are polarized along a given direction in space where the optical axis is pointing. We denote this by A, which stands for "parallel to the transmission axis." The other possible quantum state is the state in which they are both polarized along a direction "perpendicular to the first transmission axis." We denote this second quantum state by the letter E.

The quantum superposition principle allows the formation of a quantum state that contains equal amounts of the parallel polarized state and the perpendicularly polarized state. For example, if we insert in the paths of the photons crystals, with both transmission axes straight up, this will result in either both photons being in state A or both being in state E. In other words, there is a probability of one-half, or 50%, that both photons will pass through along channel A, and a probability of one-half that both will pass through channel E. In this case we have strict correlation in the outcomes of the experiments involving the two photons. Denoting one photon that flies to the "left" as the left photon and the other as the "right" photon, two typical synchronized sequences of measurements of polarization, where A stands

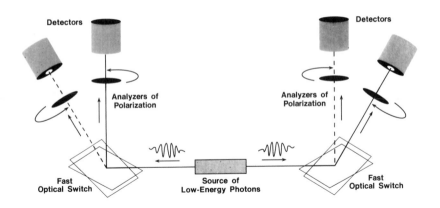

FIGURE 18. The experimental set-up carried out by Aspect and his co-workers to test the predictions of quantum theory versus the predictions of local realistic theories. Thus there is now general agreement that the experiments testing Bell's theorem have made local realistic theories, like deterministic hidden variables, scientifically gratuitous at best.

for the photon polarized along the axis or which is parallel to the optical axis, and E for the photon that is polarized perpendicular to the axis of the crystal, would look like this:

LEFT: A E A E A A E A E E E A A A

RIGHT: A E A E A A E A E E E A A A

(1)

Since the actual orientation in space of the optical axis is immaterial, it does not matter which direction in space the two optical axes point. As long as both are parallel, we could change the orientation of the axes, and the records would still look similar to the one shown in (1). One can keep track of the two optical axes of the crystals by constructing dials which read a direction in space like the hand of a clock. If both optical axes are at any angle, say along the 12:00 direction, 2:00 direction, 7:00 direction, and so on, the measurement records in all these cases will be similar to those in (1). The word "similar" is important here because any finite number of measurements will not necessarily look identical to (1). If, however, a large number of measurements are made, quantum probability predicts that 50% of the time both left and right will record an A polarization, and that 50% of the time they will both record an E polarization. Given a sufficient number of measurements, we should discover, in other words, that 50% of the time both photons are polarized along the given direction, and that 50% of the time they will be perpendicularly polarized to the given direction.

Suppose we force the optical axis of the left crystal to be, say, along the 12:00 direction, and put that on the right at 90 degrees, or at the 3:00 direction. The sequences of measurement will now look like:

$$\text{LEFT: A E A A E E A E A A A E E A E E}$$

$$(2)$$

$$\text{RIGHT: E A E E A A E A E E E A A E A A}$$

This means we have perfect anti-correlation. When the left photon passes through the 12:00 crystal and is recorded by the A detector it had a polarization parallel to it, and when it got recorded by the E detector, it had a polarization along the 3:00 direction. The opposite is true of the right photon. Since we go from the perfect matching of the sequences (1) when both axes are along the same direction to the perfect mismatching of the sequences (2) when one axis is perpendicular to the other, there must be intermediate orientations in the two directions where we don't have either perfect matchings or perfect mismatchings. In particular, there must be an intermediate angle between the two orientations for which there are three matches out of four and one mismatch out of four. The sequence of measurements will then look like this with the mismatches underlined:

$$\text{LEFT: A E } \underline{\text{E}} \text{ } \underline{\text{E}} \text{ } \underline{\text{A}} \text{ A E E A } \underline{\text{E}} \text{ A A E A A } \underline{\text{E}}$$

$$(3)$$

$$\text{RIGHT: A E } \underline{\text{A}} \text{ E } \underline{\text{E}} \text{ A E E A } \underline{\text{A}} \text{ E E A E E } \underline{\text{A}}$$

Quantum theory actually says that the angle between the two orientations will be 30 degrees (since the square of the cosine of 30 degrees is 0.75 or 3/4). If the left crystal axis is along the 12:00 direction, the right axis will have to be placed along the 1:00 direction. If the left crystal axis is along the 3:00 direction, the right axis will have to be placed along the 4:00 direction, and so on.

Finally, there must be another angle between the two orientations for which there are three mismatchings out of four, and one matching out of four. The sequence of measurements will then look like this with the mismatches underlined once again:

$$\text{LEFT: A E } \underline{\text{E}} \text{ } \underline{\text{E}} \text{ } \underline{\text{A}} \text{ } \underline{\text{A}} \text{ } \underline{\text{A}} \text{ E A E A A } \underline{\text{E}} \text{ A E } \underline{\text{E}}$$

$$(4)$$

$$\text{RIGHT: A } \underline{\text{A}} \text{ E } \underline{\text{E}} \text{ } \underline{\text{E}} \text{ } \underline{\text{E}} \text{ } \underline{\text{A}} \text{ A A } \underline{\text{A}} \text{ E } \underline{\text{A}} \text{ E } \underline{\text{A}} \text{ } \underline{\text{A}}$$

Quantum theory predicts that the angle between the two orientations is 60 degrees. If the left axis is along the 12:00 direction, then the right axis would be along the 2:00 direction. To summarize, quantum theory predicts the sequences (1), (2), (3), and (4) for the four angles between the two axes equal to 0, 90, 30, and 60 degrees, respectively. What the actual experiments testing Bell's theorem carried out in the laboratory have shown is that the predictions of quantum theory are valid, or that Bell's inequality is violated. (A simplified version of an experiment testing Bell's inequality is illustrated in Plate II, p. xi.) The results do not agree with theories consistent with Einstein's position for the simple reason that classical physics, and the theory of relativity, obey the locality assumption. Since it was inconceivable to Einstein that this assumption would not hold in experiments in which quantum theory predicts correlations over space-like separated regions, he was utterly convinced that the predictions of quantum theory would fail in this situation. The results of the experiments testing Bell's theorem have proven Einstein wrong on both counts. Other physicists, most notably David Bohm and Louis de Broglie, have sought to undermine the completeness of quantum theory with the assumption that the wave function does not provide a complete description of the system. If this were, in fact, the case, then one could avoid the conclusion in the Copenhagen Interpretation that quantum indeterminacy and probability are inescapable aspects of the quantum world, and assume that all properties of a quantum system can be known *in principle, if not in practice*. What these physicists have attempted to do is assign a determinacy at an unspecified subquantum level. They speculate that a number of variables exist on this level which are inaccessible to the observer at both the macro and quantum levels. These so-called hidden variables supposedly make a quantum system completely deterministic at that subquantum level.

In other words, these theorists argue that although quantum uncertainty or indeterminacy is apparent in the quantum domain, determinism could allegedly reign supreme at this hidden level. If deterministic hidden variables exist, then quantum indeterminacy is not absolute, and what this indeterminacy reflects is merely ignorance on the part of the scientist of the values of the hidden variables. This means that although quantum indeterminacy may be a property of a quantum system *in practice*, it need not be so *in principle*. This strategy allows one to view physical attributes of quantum systems, such as spin and polarization, as objective or "real," even in the absence of measurement, and to assume, as Einstein did, a one-to-one correspondence between every element of the physical theory and the physical reality.

Yet these so-called local realistic classical theories predict a totally different result for the correlations between the two photons in experiments testing Bell's theorem. Although the assumption that the variables are hidden or unknown will obviously not allow us to determine whether what happens at the left filter is causally connected to what happens at the right filter, we can test the reasonableness of hidden variable theory here with a simple assumption. If locality holds, or if no signal can travel faster than light, turning the right filter can change only the right sequence, and turning the left filter can only change the left sequence. According to hidden variable theories, therefore, turning the second axis from the 12:00 direction to the 1:00 direction should yield one miss out of four in the right sequence. And turning the first axis from the 12:00 direction to the 11:00 direction should yield one miss out of four in the left sequence. If we take into account the overlaps in the mismatches between the two sequences, we could conclude that the overall mismatching rate between the two sequences is *two or less out of four*. Local realistic theories, or hidden variable theories would, therefore, predict the following sequences of measurements:

LEFT: A E A A E A E E E A A A A E A E

                                                                    (5)

RIGHT: A A E A E E E E A A A E E E E E

It is clear when one compares (4) with (5) that for certain angles local realistic theories would predict records which differ significantly in their statistics from what quantum theory predicts. Bell's theorem recognizes this fact. The specific way local realistic theories differ from quantum theory is given by various kinds of Bell inequalities, and it is clear that quantum theory strongly violates such inequalities for certain angles, such as 60 degrees in the example presented here. What these results clearly suggest is that local hidden variables are not present, and that the claim that they might exist appears to be false.

Experiments Testing Bell's Theorem

The first tests of Bell's inequality were conducted at the University of California, Berkeley, and the results were reported in 1972. In the earliest tests, photons were emitted from calcium or mercury atoms which were excited into a specific energetic state by laser light. The return to the ground state from the excited state in these atoms involves an electron in two transitions between an intermediate state to the ground state, and the creation in each transition of a photon. The two photons were produced for the transitions chosen with correlated polarizations. Using photon counters placed behind polarizing filters, the photons from the cascade were then analyzed. In the mid 1970s experiments were conducted in which the photons were gamma rays produced when an electron and a positron annihilate, and the polarizations of the two photons were correlated. In the many tests that have been conducted since the mid 1970s the impulse has been to eliminate any problems in the design of earlier experiments, and to insure that the detectors are placed far enough apart so that no signal traveling at light speed can be assumed to be accounting for the correlations.

Although these experiments produced results which were in accord with the predictions of quantum theory, or which violated Bell's inequality, it was still possible to assume that the wave function at the source of the two-photon systems was the carrier of the information about the experimental system. This assumption allowed one to avoid confronting the prospect that the correlations violated locality, or occurred faster than the time required for light to carry signals between the two regions. Although the notion of a single wave function carrying information over any distance from one source as if it were one wave function was a bit strange, it was at least a basis for denying the existence of non-locality. What was needed to dispel this notion was an experimental arrangement in which the structure of the experiment could be changed when the photons were in flight from their original source in analogy with Wheeler's delayed-choice experiment. It was an arrangement of this sort which was the basis for the experiments conducted at the Institute of Optics at the University of Paris at Orsay. The results, published in 1982, provided unequivocal evidence that Einstein's view of realism, which had been extended into local hidden variable theories or local realistic theories, is incompatible with quantum physics.

In these experiments, conducted by Alain Aspect and his colleagues Jean Dalibard and Gerard Roger, the choice between the orientations of the polarization analyzers is made by optical switches while the photons are flying away from each other.[6] The beam can be directed toward either one of two polarizing filters which measure a different direction of polarization, and each has its own photon detector behind it. The switching between the two different orientations takes only 10 nanoseconds, or 10×10^{-9} sec, and occurs with an automatic device that generates a pseudorandom signal. Since the distance between the two filters is 13 meters, no signal traveling at the speed of light can be presumed to carry information between the filters. A light signal would take 40 nanoseconds to go from one filter to the other. This means, assuming that no signal can travel faster than light, that the choice of what orientation of polarization is measured on the right could not influence the transmission of the photon through the left filter. The results of these experiments also agree with quantum mechanical predictions of strong correlations, and Bell's inequality is violated. As the French physicist Bernard d'Espagnat puts it, "Experiments have recently been carried out that would have forced Einstein to change his conception of nature on a point he always considered essential ... we may safely say that non-separability is now one of the most certain general concepts in physics."[7]

Non-separability, or non-locality, as a fact of nature does not, incidentally, mean that we have discovered the basis for faster-than-light communications. The primary reason why this is the case is that there is no way to establish causal links that carry useful information between paired particles in this situation. The effect that is studied in the modern EPR-like experiments applies only to events that have a common origin in a unified quantum system, like the annihilation of a positron-electron pair, the return of an electron to its ground state, or the separation of a pair of photons from the singlet state. Since any information that originates from these sources is, in accordance with quantum theory, a result of quantum indeterminacy, the individual signals are random. In the case of light, the polarizations, or spins, of each of the photons in the Aspect experiments carry no information, and any observer of the photons transmitted along a particular axis would see only a random pattern. This pattern makes non-random sense only if we are able to compare it with the pattern observed in the other paired photon. Any information contained in the paired photons derives from the fact that the properties of the two photons exist in complementary relation, and that information is uncovered only through a comparison of the difference between the two random patterns.

Confronting a New Fact of Nature

Although the discovery that non-locality is a fact of nature will not provide the basis for a technological revolution in the telecommunications industry, it does represent a rather startling new addition to our scientific world-view. And it is also likely to occasion some massive readjustments in our understanding of both the character and the ultimate foundations of the knowledge we call physics. If we can also assume, as the historical record suggests that we should, that our scientific world-view plays a large role in conditioning our understanding of the character of self and world, non-locality as a new fact of nature could be just as revolutionary in these terms as the confirmation of Copernican hypothesis in the seventeenth century. As Henry Stapp puts it, non-locality could be the "most profound discovery in all of science."[8]

As Bernard d'Espagnat has argued,[9] non-locality should obviously not be assumed to be a fact of nature only in the special laboratory conditions in the experiments testing Bell's theorem. What these experiments reveal is a general property of nature in which particles that interact with other particles in these terms must be viewed as a single quantum system which responds together to further interactions. Virtually everything in our immediate physical environment is made up of quanta that have been interacting with other quanta in this manner from the Big-Bang to the present. The atoms in our bodies are made up of particles that were once in close proximity to the cosmic fireball, and other particles that interacted at that time in a single quantum state can be found in the most distant star. This means, however strange or bizarre it might seem, that the quanta that make up our bodies are as much a part of a unified system as the photons propagating in opposite directions in the Aspect experiments. Thus non-locality, or non-separability, in these experiments translates into the vastly grander notion of non-locality, or non-separability, as factual condition in the entire universe. Although this clearly argues for holism, as d'Espagnat and Bohm have pointed out, there is, as we hope to demonstrate, a good deal more implied here than the conclusion that "one is all" or "all is one." Equally significant, in our view, is what is suggested about the limits of scientific knowledge.

Although there is little doubt among physicists that non-locality must now be recognized as a fact of nature, not much has been done thus far to explore the larger implications beyond the somewhat grudging conclusion that the Copenhagen Interpretation of quantum mechanics must remain the "orthodox" interpretation. We have made it our business here to explore the implications of this fact of nature not only in terms of what it means for scientific epistemology, or for our scientific world-view generally. We also intend to explore its meaning in terms of what this fact, along with the totality of other facts in modern physics, implies about the character of human consciousness, particularly as it pertains to a new understanding of a fundamental relationship—that between the part and whole as it has been disclosed in physical theories—since the special theory in 1905. We will begin this admittedly ambitious journey in the next chapter by demonstrating that the principle of complementarity as it was defined by Niels Bohr has not been well understood among the community of physicists, and that a better understanding serves to resolve many of the seeming paradoxes associated with the quantum mechanical description of nature.

consciousness does not involve transfer of energy or mass to another ------ so, non-separability or connectedness does Not apply here.

4
Changing the Rules:
The New Epistemology of Science

I am afraid of this word Reality.
—*Arthur Eddington*

The most fundamental aspect of the western intellectual tradition, which is pervasive throughout the history of western thought, is, as virtually anyone who has seriously studied this history knows very well, ontological dualism. The concept of Being as continuous, immutable, and having a prior or separate existence from the world of change dates from Parmenides. The same qualities were associated with the god of the Judeo-Christian tradition, and were considerably amplified by the role played in theology by Platonic and Neoplatonic philosophy. Since the architects of classical physics were all inheritors of a cultural tradition in which ontological dualism was a primary article of faith, the idealization of the mathematical ideal as a source of communion with God, which dates from Pythagoras, provided an ontological or metaphysical foundation for the emerging natural sciences. As we will see in more detail in the next chapter, doing physics for the architects of classical physics was clearly a form of communion with the geometrical and mathematical forms resident in the perfect mind of God. *→ God and Mind do not exist*

The role of seventeenth century metaphysics is also apparent in a less obvious way in metaphysical presuppositions about the character of the matter described by classical equations of motion. These presuppositions can be briefly defined as follows: 1) the physical world is made up of inert and changeless matter, and matter changes only in terms of location in space; 2) matter mirrors physical theory in that its behavior is inherently mathematical, and thus physical reality is essentially quantitative rather than qualitative; 3) matter as the unchanging unit of physical reality can be exhaustively understood by mechanics, or by the applied mathematics of motion; and 4) the mind of the observer is separate from the observed system of matter, and the ontological bridge between the two is physical law and theory.[1]

All of these presuppositions have a metaphysical basis in that they are required in order to assume that the full and certain truths about the physical world are revealed in a mathematical structure governed by physical laws which have a prior or separate existence. Although Kepler, Galileo, Descartes, and Newton assumed that the metaphysical or ontological foundation for these laws was the perfect mind of God, that assumption would be increasingly regarded, even in the eighteenth century, as ad hoc and unnecessary for the reasons noted earlier. What would endure in an increasingly disguised form was the assumption of ontological dualism which allowed the truths of mathematical physics to be regarded as having a separate and immutable existence outside the world of change.

Although metaphysical presuppositions about the character of matter would hold through most of the nineteenth century, any overt appeal to the metaphysics soon became, as we have seen, unfashionable. The science of mechanics was increasingly regarded, suggests Ivor Leclerc, as "an autonomous science," and any alleged role of God as "deus ex machina."[2] At the beginning of the nineteenth century, Laplace, along with a number of other great French mathematicians, advanced the view that the science of mechanics constituted a "complete" view of nature. Since this science, by observing its epistemology, had revealed itself to be the "fundamental" science, the hypothesis of God was, they concluded, entirely obtuse and unnecessary.

Laplace came to be recognized not only for eliminating the theological component of classical physics but the "entire metaphysical component" as well.[3] The epistemology of science requires, he said, that we proceed by inductive generalizations from observed facts to hypotheses "tested by observed conformity of the phenomena to the hypotheses."[4] In contrast with Newton, who insisted on the avoidance of hypotheses, Laplace accepted hypotheses as a means of correcting the study of phenomena for the discovery of general laws. What was unique in Laplace's view of hypotheses was his insistence that we cannot attribute reality to them. Although concepts like force, mass, motion, cause, and laws are obviously present in classical physics, they exist in Laplace's view only as "quantities." Physics is concerned, he argued, with quantities which we associate as a matter of convenience with concepts, and the "truths" about nature are only the quantities.

As this view of hypotheses and the truths of nature as quantities was extended in the nineteenth century to a mathematical description of phenomena like heat, light, electricity, and magnetism, Laplace's assumptions about the actual character of scientific truths seemed quite correct. This progress suggested that if we could remove all thoughts about the "nature of" or the "source of" phenomena, or if we could eliminate all intuitive conceptions of the processes evinced by observed phenomena, then the pursuit of strictly quantitative concepts would bring us to a complete description of all aspects of physical reality. Subsequently, figures like Mach, Kirchoff, Hertz, and Poincare developed a program for the study of nature that was quite different from that of the architects of classical physics.[5]

The seventeenth century view of physics as a "philosophy of nature" or as "natural philosophy" was, therefore, displaced by the view of physics as an autonomous science which was "the science of nature."[6] The new program promised to subsume all of nature by a mathematical analysis of entities in motion. The knowledge that is physics, or the "true" understanding of nature, was the mathematical description. This doctrine, known as positivism, states, once again for emphasis, that true, genuine, and certain knowledge is revealed in the mathematical description, and thus all metaphysical concerns are, by definition, excluded. Although there was, as we shall see, considerable ambiguity in Einstein's view of positivism, he was quite consciously following Mach's conception of this program when he developed both the special and general theories.

The first major blow to the notion that mathematical physics discloses the full and certain truths about physical reality came with the discovery of non-Euclidean geometry in the early nineteenth century. That it was possible to conceive of mathematically self-consistent geometries that were quite different from the geometry that mathematical physics had previously alleged to be one of the full and certain truths of nature was unsettling. The suggestion that there was an element of "subjectivism" in the creation of mathematical structures was explored by Kant. He argued that our earlier assumption that knowledge of the world in mathematical physics is wholly determined by the behavior of physical reality could well be false. Perhaps, said Kant, the reverse is true—that the objects of nature conform to our knowledge of nature. The relevance of the Kantian position was later affirmed by the leader of the Berlin school of mathematics, Wierstrass, who came to a conclusion that would also be adopted by Einstein—that mathematics is a pure creation of the human mind.[7]

The lively debate over the epistemological problems presented by quantum physics, reflected in the debate between Einstein and Bohr, came, as the physicist and historian of science Gerald Holton has demonstrated, to a grinding halt shortly after World War II. What seems to have occurred, as Holton understands it, is that the position of Einstein became the accepted methodology in contemporary research.[8] Yet that position, as Leclerc explains, contains some fundamental ambiguities.[9] Although Einstein was in full agreement with the notion that physical theories are the free invention of the human mind, he, nevertheless, also maintained that "the empirical contents of their mutual relations must find their representations in the conclusions of the theory."[10] Einstein sought to reconcile the fundamental ambiguity between the two positions, that physical theories "represent" empirical facts, and that physical theories are a free invention of the human intellect, with an article of faith. "I am convinced," wrote Einstein, "that we can discover by means of purely mathematical constructions the concepts and laws connecting them with each other, which furnish the key to understanding natural phenomena."[11] Since the lack of a one-to-one correspondence between every element of the physical theory and the physical reality in quantum physics completely undermines this conviction, how does Einstein sustain it? He does so, suggests Leclerc, by appealing to "a tacit seventeenth century presupposition of metaphysical dualism and a doctrine of the world as a mathematical structure completely knowable by mathematics."[12]

It is essentially this position that underlies the methodology of physics after World War II. The tacit presupposition and the doctrine this presupposition serves to protect have, for reasons we will explore more fully later, enormous psychological and emotional appeal for the trained physicists. Since physicists could reasonably assume that positivism had purged scientific knowledge of all vestiges of the metaphysical, there seems to have been little active awareness that metaphysical dualism and the doctrine of the world as completely knowable by mathematics were being implicitly appealed to in the conduct of physics. It is this background, as we noted in the beginning of this discussion, that allows us to appreciate why many theoretical physicists have been enticed into making what can fairly be called metaphysical leaps in their efforts to save classical assumptions in their dealings with the character of quantum mechanical reality. Since Bell's theorem and the experiments testing that theorem force us to evaluate these assumptions within the normal conduct of science, we are now able to disclose that these assumptions derive, unwittingly to be sure, from metaphysical presuppositions that were not previously viewed as such because they could be construed as "self-evident" prior to these developments in the normal conduct of science. Yet in order to disclose the role played by these metaphysical presuppositions in the alternatives to the Copenhagen Interpretation (CI), or in the "quantum ontologies" discussed in chapter 7, we must first appreciate why CI "must" be invoked in our dealings with quantum mechanical reality, and the central and vitally important role played in this interpretation by the principle of complementarity.

The Dilemma of Quantum Epistemology

We do not mean to imply, of course, that the community of physicists has been unaware of the threats posed by quantum physics to these assumptions. Quantum physics profoundly disturbed physicists from its very inception because quantum mechanical experiments yield results, as we have seen, that are clearly "dependent" upon observation and measurement, and where a one-to-one correspondence between every element of the physical theory and the physical reality cannot be confirmed. For this reason physicists have also been obliged to pay grudging homage to CI as the most commonly accepted understanding of the epistemological situation in quantum physics. Yet physicists have, by and large, been willing to accept the "orthodoxy" of this interpretation only if they could also assume that the foundational principles involved need not apply to all of physics, or that advances in physical theory, like local realistic theories, might eventually displace them altogether. What many physicists have found most unsettling about the results of experiments testing Bell's theory is that they seem to make both of these prospects quite unrealistic.

The central pillar of CI is the principle of complementarity, most eloquently developed by Bohr. The usual textbook definition of complementarity suggests that it applies to "apparently" incompatible constructs, like wave and particle, or variables, such as position and momentum. Since one of the paired constructs or variables cannot define the situation in the quantum world in the absence of the other, both are required for a complete view of the actual physical situation. Thus a description of nature in this "special" case requires that the paired constructs or variables be viewed as complementary, meaning that both constitute a complete view of the situation while only one can be applied in a given situation. The textbook definition normally concludes with the passing comment that since the experimental situation determines which complementary construct or variable will be displayed, CI assumes that entities in the quantum world, like electrons or photons, do not have definite properties apart from our observation of them.

One obvious reason why complementarity is dealt with in such a cursory and inadequate manner in most physics textbooks is that the writers of the textbooks have been able to assume until recently that CI either does not apply to all of physics, or that it can be viewed as a provisional and passing interpretation. Yet another reason could be that Bohr's efforts to achieve the utmost clarity often resulted in a prose so riddled with qualifications that it is difficult to determine his precise meaning. When we examine his statements in the light of recent developments in physics, however, it is not difficult to see how precise they really are.

Much of the confusion about Bohr's understanding of the epistemological situation in quantum physics seems to derive from his frequent description of quantum mechanics as a "rational generalization of classical mechanics" and his requirement that the results of quantum mechanical experiments "must be expressed in classical terms."[13] As the physicist and philosopher of science Clifford Hooker suggests, when one reads these statements out of context, one could conclude that quantum mechanics is an "extension" of classical mechanics, and that our experience in the quantum domain is merely a special case in which working hypotheses and assumptions from classical mechanics must be modified while remaining fundamentally unchallenged.[14] Reading the statements out of context has also led many physicists to wrongly presume, as physicists Herman Feshbach and Victor F. Weisskopf have recently emphasized[15], that Bohr suggests that the measuring instruments follow classical physics rather than quantum physics.

When we look at Bohr's statements in context, however, we discover that he viewed classical mechanics as a subset of quantum mechanics, or as an "approximation" which has a limited domain of validity. Quantum mechanics, concluded Bohr, is the complete description, and the measuring instruments in quantum mechanical experiments obey this description. Although we can safely ignore quantum mechanical effects in dealing with macro-level phenomena for the reason that those effects are small enough to be excluded for "practical" purposes, we cannot ignore the implications of quantum mechanics on the macro level for the obvious reason that they are there. Since the quantum of action is always present on the macro level, it entails, said Bohr, "a final renunciation of the classical ideal of causality and a radical revision of our attitude toward the problem of physical reality."[16] Since classical causality presumes a one-to-one correspondence between every element of the physical theory and the physical reality, and since the quantum of action obviates this correspondence on the macro level as well, Bohr concluded, long before this became dogma among physicists, that unrestricted causality simply does not exist in nature.

In classical physics quantities like position and momentum, and laws like energy-momentum conservation, can be simultaneously applied in a single unique circumstance. Thus the results of experiments are precisely those which are predicted in physical theory. In quantum physics, however, such constructs are complementary, or mutually exclusive in accordance with the indeterminacy principle. This means, says Bohr, that the "fundamental postulate of the quantum of action ... forces us to adopt a new mode of description designated as complementary in the sense that any given application of classical concepts precludes the simultaneous use of other classical concepts which in a different connection are equally necessary for the elucidation of phenomena."[17]

Since the principle of complementarity will assume increasingly more importance in the remainder of this discussion, let us pause for a moment and consider why there has been a tendency to ignore its implications for all of physics. In dealing with the behavior of macro-level objects, the smallness of the quantum of action compared to macroscopic values is such that we do not need to use quantum mechanics to get reliable results. Quantum indeterminacy in a flying tennis ball is, for example, exceedingly small due to the smallness of the de Broglie wavelength, which is less than one-hundred-thousandth of a billionth of a billionth of a billionth of a centimeter. Thus the deterministic equations of classical physics are more than adequate for predicting how the ball will fly through the air. The initial impact of the racket "causes" the ball to move in a particular direction with a particular speed, or momentum, and its subsequent motion in space seems utterly predictable. If we take care to factor in all the initial macro-level conditions, the ball seems to appear precisely where we predicted it would. There is no reason to assume that our observations of the ball have had any impact whatsoever on these results. And it would seem rather insane to imagine that the ball might not appear precisely where it did had we chosen not to observe it. Our effort to coordinate experience with physical reality on the tennis court suggests that this reality is utterly deterministic. And the same applies to every object that we are capable of manipulating in normative experience.

Yet, as Bohr realized, when we apply classical mechanics on the tennis court, or anywhere else in dealing with objects on the macro level, we are being subjected to a macro-level illusion. As Hooker puts it, "Bohr often emphasizes that our descriptive apparatus is dominated by the character of our visual experience and that the breakdown in the classical description of reality observed in relativistic and quantum phenomena occurs precisely because we are in these two regions moving out of the range of normal visualizable experience."[18] Although our experience with macro-level objects bears no resemblance to our experience with quantum particles, those objects come into existence as a result of interaction between fields and quanta. Unrestricted causality could be assumed to exist in nature as long as it was possible to presume that all the initial conditions in an isolatable system could be completely defined, and that every aspect of this system corresponds with every element of the physical theory which describes it. Yet the quanta that make up macro-level systems cannot be said to have definite properties in the absence of observation—between observations they can be in some sense, as Richard Feynman suggests, "anywhere they want" within the prescribed limits of the uncertainty principle.

When Bohr says that the quantum of action "forces" us to adopt a new "mode" of description, he is not suggesting, as Einstein derisively commented, that the moon is not there when it is not being observed.[19] Bohr is simply describing a new epistemological situation which we are "forced" to accept now that we realize that the quantum of action is, like light speed and the gravitational constant, a constant of nature. If this were not so, classical causality, and classical determinism would remain firmly in place. Since the quantum of action is a constant of nature, adopting a new mode of description is not, as Bohr's colleague Leon Rosenfeld notes, "something that depends on any free choice, about which we can have this or that opinion. It is a problem which is imposed upon us by Nature."[20] The situation is comparable, said Bohr to that which we faced earlier in coming to terms with the implications of relativity theory:

> The very nature of quantum theory thus forces us to regard the space-time coordination and the claim of causality, the union of which characterizes the classical theories, as complementary but exclusive features of the description, symbolizing the idealizations of observation and definition respectively. Just as relativity theory has taught us that the convenience of distinguishing sharply between space and time rests solely on the smallness of velocities ordinarily met with compared to the speed of light, we learn from the quantum theory that the appropriateness of our visual space-time descriptions depends entirely on the small value of the quantum of action compared to the actions involved in ordinary sense perception. Indeed, in the description of atomic phenomena, the quantum postulate presents us with the task of developing a "complementary" theory the consistency of which can be judged only by weighing the possibilities of definition and observation.[21]

Just as we can safely disregard light speed in our application of classical dynamics on the macro level because the speed of light is so large that relativistic effects are negligible, so can we disregard the quantum of action on the micro level because its effects are so small. Yet everything we deal with on the macro level obeys the rules of relativity theory and quantum mechanics. Classical physics is a workable approximation which seems precise only because the largeness of the speed of light and the smallness of the quantum of action give rise to negligible effects. The notion from classical physics that the observer and the observed system are separate and distinct is also, Bohr suggests, undermined by relativity theory before it is undermined in a slightly different way by quantum physics. Just as one cannot in relativity theory view the observer as outside the observed system, in that one must assign that observer particular space-time coordinates relative to the entire system, so one must view the observer in quantum physics as an integral part of the observed system. There is in both cases no "outside" perspective.

Another observation made by Bohr here, which will assume more importance later in this discussion, is that space-time coordination and the claim of causality are complementary constructs which "symbolize" the idealizations of observation and definition respectively. What Bohr is obviously assuming is that the more complete description of nature is the quantum mechanical description and, therefore, that relativity theory is grounded in an outmoded classical framework. When we apply the logical framework of the more complete description to relativity theory, it is then possible, he suggests, to disclose the normally unexamined idealizations embedded in that theory. Bohr then goes on to suggest that an improved understanding of definition and observation will be derived from advances in quantum physics. At the same time relativity theory did disclose, according to Bohr, a profound new complentarity in the equivalence between mass and energy — each excludes the other in any given situation, and yet both constructs are necessary for a complete understanding of the entire physical situation.

Realism Versus Idealism in the Quantum World

The power of Bohr's arguments derives largely from his determination to remain an uncompromising realist in the sense of insisting that all conclusions be consistent with experimental conditions and results, and refusing to make any metaphysical leaps. He had enormous and unfailing respect for the stern gatekeeper which has habitually stood at the door of scientific knowledge—measurement or observation under controlled and repeatable experimental conditions is necessary to confirm the validity of any scientific theory. What we know about phenomena as a result of the experiments confirming the validity of quantum physics refers exclusively, said Bohr, to the "observations and measurements obtained under specific circumstances, including an account of the whole experimental arrangement."[22]

Bohr concludes that if we view phenomena in this way, we cannot conceive of the act of observation or measurement as "disturbing phenomena ... or creating physical attributes of atomic objects."[23] We can assume that we "disturb" or "create" phenomena via observation or measurement only if we make the prior assumption that the atomic world is describable "independent" of observation and measurement. As Hooker puts it, "There is no 'disturbance' here in the classical sense of a change of properties from one as yet unknown value of some autonomously possessed physical magnitude to a distinct value of that magnitude under the causal action of the measuring instrument. Even talk of change of properties, or creation of properties, is logically out of place here because it presupposes some autonomously existing atomic world which is describable independently of our experimental investigation of it."[24] The hard lesson here from the point of view of classical epistemology is that there is no godlike perspective from which we can know physical reality "absolutely in itself." What we have instead is a mathematical formalism through which we seek to unify experimental arrangements and descriptions of results.

"The critical point," said Bohr, "is here the recognition that any attempt to analyze, in the customary way of physics, the 'individuality' of atomic processes, as conditioned by the quantum of action, will be frustrated by the unavoidable interaction between the atomic objects concerned and the measuring instruments indispensable for that purpose."[25] Although we are doing what we have always done in physics, setting up well-defined experiments and reporting well-defined results, the difference is that any systematized, definite statements about results must include us and our measuring apparatus. Since the quantum of action is unavoidable present, a one-to-one correspondence between the categories associated with the complete theory and the quantum system can never be reflected in those results. For this reason, concludes Bohr, "radiation in free space as well as isolated material particles are abstractions, their properties being definable and observable only through their interactions with other systems."[26] When we use classical terms to describe the state of the quantum system, we simply cannot assume that the system possesses properties which are independent of the act of observation of them. We can make that assumption only in the absence of observation.

What is dramatically different about this new situation is that we are "forced" to recognize that our knowledge of the physical system cannot in principle be complete or total. Although we have in quantum mechanics complementary constructs which describe the entire situation, the experimental situation precludes simultaneous application of complementary aspects of the complete description. The choice of which is applied is inevitably part of the results we get. Although the conceptual context of our descriptions remains classical and does not change, we are obliged to use a new logical framework based on a new epistemological foundation to make sense out of the observed results.

Complementarity and Objectivity

Before we discuss in more detail what Bohr means by complementarity, we should dispense with another large misunderstanding of his position. Some have assumed that because Bohr's analysis of the conditions for observation precludes exact correspondence between every element of the physical theory and the physical reality, that he is in some sense implying that the reality does not objectively exist, or that we have ceased to be objective observers of this reality. These conclusions are possible only if we equate physical reality with our ability to know it in an absolute sense. Does nature become real when we, like the God of Bishop Berkeley, have absolute knowledge of its character, or does it cease to be real when we discover that we lack this knowledge? Bohr though not:

> The notion of complementarity does in no way involve a departure from our position as objective observers of nature, but must be regarded as the logical extension of our situation as regards objective description in this field of our experience. The recognition of the interaction between the measuring tools and the physical systems under investigation has not only revealed an unsuspected limitation of the mechanical conception of nature, as characterized by attribution of separate properties to physical systems, but has forced us, in ordering our experience, to pay proper attention to the conditions of observation.[27]

In paying proper attention to the conditions of observation, we are forced to abandon the mechanistic or classical concept of causality, and, consequently, the assumption that scientific knowledge can be complete in the classical sense. But it certainly does not follow that we have ceased to be objective observers of physical reality, or that we cannot affirm the existence of that reality. It is rather that the requirement to be objective has led us in our ongoing dialogue with nature to a new logical framework for objective scientific knowledge which Bohr labeled complementarity.

This new logical framework, said Bohr, "points to the logical condition for description and comprehension of experience in quantum physics."[28] Although normally referred to as the "principle of complementarity," the use of the word "principle" is unfortunate in that complementarity is not a principle as that word is used in physics. Complementarity is rather a "logical framework" for the acquisition and comprehension of scientific knowledge which discloses a new relationship between physical theory and physical reality which effectively undermines all appeals to metaphysics.

The logical conditions for description can be briefly summarized as follows: In quantum mechanics, the two conceptual components of classical causality, space-time description and energy-momentum conservation, are mutually exclusive, and can only be coordinated through the limitations imposed by Heisenberg's indeterminacy principle. The more we know about position the less we know about momentum and vice versa. "Contradiction," as Rosenfeld explains, "arises when one tries to apply both of them to the same situation, irrespective of the circumstances of the situation.... However, if one reflects on the use of all physical concepts, one soon realizes that any such concept can be used only within a limited domain of validity."[29]

The logical framework of complementarity, which can be applied to understanding the character and limits of all modern physical theories, is useful and necessary when the following requirements are met: 1) when the theory consists of two individually complete constructs; 2) when the constructs preclude one another in a description of the unique physical situation to which they both apply; and 3) when both constitute a complete description of that situation. Whenever we discover a situation in which complementarity clearly applies, we necessarily confront an imposing limit to our knowledge of this situation. Knowledge here can never be complete in the classical sense because we are unable to simultaneously apply the mutually exclusive constructs which constitute the complete description. The list of those situations, as we will suggest later, is much longer than Bohr could have imagined, and we speculate that it will become even longer with the advance of scientific knowledge.

When Bohr first suggested that we live in a quantum mechanical universe in which classical mechanics appears complete only because the effects of light speed and the quantum of action are negligible in arriving at useful results, one could still argue, as Einstein did, that quantum indeterminacy would be circumvented by a more complete theory. That has not happened, and there are no suggestions that it will ever happen. If quantum physics is as rock bottom in its understanding of the dynamics of physical phenomena as it now appears to be, the new situation disclosed in quantum physics should not be relegated to the "special" case of experiments in this physics. It should apply to the entire body of knowledge we call physics with consequences, as Bohr fully appreciated, that are quite imposing. "The notion of an ultimate subject as well as conceptions of realism and idealism," wrote Bohr, "finds no place in objective description as we have defined it."[30] This means that physical laws and theories do not have, as the architects of classical physics supposed, an independent existence from ourselves. They are human products with a human history useful to the extent that they help us coordinate a greater range of experience with nature. "It is wrong," said Bohr, "to think that the task of physics is to find out how nature is. Physics concerns what we can say about nature."[31]

The Necessity to Use Classical Concepts

Why, then, did Bohr stipulate that we must use classical descriptive categories, like space-time description and energy-momentum conservation, in our descriptions of quantum events? If classical mechanics is an approximation of the actual physical situation, it would seem to follow that the classical descriptive categories are not adequate to describe this situation. If, for example, quantities like position and momentum are "abstractions" with properties that are "definable and observable only through their interactions with other systems," then why should we represent these classical categories "as if" they were actual quantities in physical theory and experiment? Although Bohr's rationale for continued reliance on these categories is rarely discussed, it carries some formidable implications for the future of scientific thought. The rationale is based upon an understanding of the manner in which scientific knowledge discloses the subjective character of human reality:

> As a matter of course, all new experience makes its appearance within the frame of our customary points of view and forms of perception. The relative prominence accorded to the various aspects of scientific inquiry depends upon the nature of the matter under investigation ... occasionally ... the [very] "objectivity" of physical observations becomes particularly suited to emphasize the subjective character of experience.[32]

The history of science grandly testifies to the manner in which scientific objectivity results in physical theories which must be assimilated into "customary points of view and forms of perception." As we engage in this assimilation process, it does occasionally happen that the subjective character of experience is emphasized in unexpected ways. The framers of classical physics derived, like the rest of us, their customary points of view and forms of experience from macro-level visualizable experience. Although some classical constructs, like action-at-a-distance, had no direct analogue in everyday visualizable experience, and to that extent emphasized the subjective character of this experience, the mathematical generalizations that explained action-at-a-distance were visualizable. Thus the descriptive apparatus of visualizable experience came to be reflected in the classical descriptive categories.

A major discontinuity appears in the progress of physics as we moved from descriptive apparatus dominated by the character of our visualizable experience to a more complete description of physical reality in relativistic and quantum physics. The actual character of physical reality in modern physics lies, as we have seen, completely outside the range of visualizable experience. Einstein, as the following passage suggests, was also acutely aware of this discontinuity: "We have forgotten what features of the world of experience caused us to frame [prescientific] concepts, and we have great difficulty in representing the world of experience to ourselves without the spectacles of the old-established conceptual interpretation. There is the further difficulty that our language is compelled to work with words which are inseparably connected with those primitive concepts."[33] Yet there is also continuity in this progress in that classical descriptive categories provided the framework for "rational generalization" that led to relativity theory and quantum mechanics.

Bohr concluded, therefore, that we must use the classical descriptive categories not because there is anything sacrosanct about them. The reason we must use them is that our ability to communicate unambiguously is bounded by our experience as macro-level perceivers. On this level the effects of light speed and the quantum of action are far too negligible to condition our normative conceptions of subjective reality. As the French philosopher Henri Bergson was among the first to point out, our logic is the logic of solid bodies, and is derived as a result of experience on the macro level. The psychologist Jean Piaget would later provide some substantive validity to Bergson's claim (1950) in his studies of the cognitive development of children. Those studies indicate that logical and mathematical operations result from the internalization of operations executed originally with solid bodies.[34] Those operations appear to be self-evident aspects of reality-in-itself on the macro level for the obvious reason that we cannot directly observe the quantum mechanical processes from which the apparent solidity of the objects manifests.

As the physicist Fritjof Capra notes, when, for example, we "see" any object, it is the properties of photons and of electromagnetism that are the physical foundations of the process through which what will become visual information impacts the retinal wall. Since the observation takes place through electromagnetic interactions, the apparent isolation of the observer from the observed system is an illusion created by the smallness of the quantum of action. The large number of photons involved in the process of seeing coupled with the smallness of the intervening photon connections disguise the discrete nature of quantum events.[35] Although we can assume that our macro-level logical and mathematical operations do not fully describe the reality of micro-level processes, the fact that we must learn them on the macro level has some operational consequences. In a curious but not incomprehensible sense we know in physical theory what we cannot say as macro-level perceivers living, as we must, in a subjective reality constructed in accordance with experience on this level. The following remark by Bohr illustrates this perspective:

> Of course it may be that when, in a thousand years, the electronic computers begin to talk, they will speak a language quite different from ours and lock us up in asylums because they cannot communicate with us. But our problem is not that we do not have adequate concepts. What we lack is a sufficient understanding of the unambiguous use of the concepts we have.[36]

What is important here is that Bohr does not foreclose on the possibility that there could be a different language for communicating scientific knowledge superior to our own. Yet he suggests that such a language could only evolve in a non-human intelligence whose awareness of the character of physical reality would presumably be conditioned "entirely" by the understanding of this reality contained in physical theory. Such an intelligence, he suggests, would not be obliged to negotiate its survival by coordinating experience in everyday language and logic based on macro-level visualizations, and would, therefore, conceive physical reality without these distorting spectacles. Bohr's comment that this intelligence might regard us as quite insane is probably not as much an attempt at black humor as it is an attempt to emphasize how different physical reality would appear to an intelligence that conceived it in the absence of these distorting spectacles. Yet we should also realize, says Bohr, that the concepts we have developed as macro-level perceivers are "adequate" for unambiguous communication providing that we properly understand them. His candidate for leading us to this improved understanding was the logical framework of complementarity.

It is safe to say that CI, particularly as Bohr described it, has occasioned more dogged resistance from scientists than any other "orthodox" interpretation in the history of scientific thought. Einstein and Schrödinger, as we saw in the discussion of the cat-in-a-box thought experiment, were early detractors, and the list of other prominent physicists who have sought in various ways to undermine CI is impressively long. Most of the detractors are identified as holding the so-called *realist* position, as opposed to the "instrumentalist" or "idealist" position of Bohr and others. The choice of the term realist is intriguing in that those who are identified as such are, like Einstein in the EPR thought experiment, forced into the position of claiming that a quantity "must" be called real within the context of physical theory even if it cannot be disclosed by observation and measurement in a single instance. In order to be a realist in these terms, one must abandon the imminently realistic scientific credo that experimental evidence is an absolute requirement for the validation of physical theory.

Bohr is sometimes termed an anti-realist by historians of science primarily because he concluded that complementary aspects of a quantum system, like wave and particle, cannot be regarded as mirroring or picturing the entire object system. Yet Bohr's conclusion follows from the utterly realistic fact that our interactions with this system preclude the appearance of both complementary aspects in particular measurement interactions. The occasional use of the term idealist in reference to Bohr's position is equally misleading in that it should apply to the so-called realists who assert the existence of an ideal system with properties that cannot be simultaneously measured. Although the term instrumentalist is marginally more appropriate, it carries associations with the term pragmatism, and suggests that there is something more "essential" here that physics will eventually disclose. If we want to put a proper label on Bohr's position, we should purge the term realism of pre-scientific associations and apply it to that position. Bohr is brutally realistic in epistemological terms.

CI and the Aspect Experiments

If we view the results of the Aspect experiments in terms of Bohr's version of CI, there is no ambiguity in the experimental results. The correlations between results at points A and B are in accordance with the predictions of quantum physics, and thus we appear to have a complete physical theory which coordinates our experience with this reality. Since indeterminacy is implicit in this theory and the results make no sense without it, this factual condition has important consequences which cannot be ignored. The logical framework of complementarity, premised on the scientific precept that measurement or observation is required to validate any physical theory, also requires that the conditions for observation be taken into account in the analysis of results. These conditions dictate that the two fundamental aspects of quantum reality, wave and particle, are complementary. Although both constructs are required for a complete view of the situation, the conditions for observation or measurement preclude the simultaneous application of both constructs.

If we insist that one view of the situation is the complete description in our analysis of results, like that implied by the manner in which the wave function is viewed in the many-worlds interpretation, then we are obliged to presume that something in A "causes" something to happen in B in accordance with the "deterministic" wave function. The resultant ambiguities are described as follows by Henry Stapp: "If one accepts the usual ideas about how information propagates through space and time, then Bell's theorem shows that the macroscopic responses cannot be independent of faraway causes. The problem is neither alleviated by saying that the response is determined by 'pure chance.' Bell's theorem proves precisely that the determination of the macroscopic response must be 'nonchance,' or at least to the extent of allowing some sort of dependence of this response on faraway causes."[37] Accepting the usual ideas about how information propagates through space and time means remaining attached to the classical concepts of locality and unrestricted causality. If we insist on this perspective and refuse to apply the logical framework of complementarity, the results of the Aspect experiments are more than ambiguous—they make no sense at all.

If we approach this situation, as Bohr says we must, with an analysis of the conditions for the experiment, it is clear that we cannot even begin to understand the correlations in the absence of the assumption of indeterminacy, and that we cannot, therefore, confirm the results in the absence of measurement. As the philosopher of science Henry Folse has observed, this means that "apart from the interactions with the detectors," the system which yields these results "exists in a single, non-analyzable quantum state." Even though our experience as macro-level perceivers entices us to picture the system in the Bell-type experiments as consisting of "spatially separated particles fleeing a common origin," complementarity indicates that this is a distorted view of the "wholeness" of the interaction in which the quantum system is prepared and which includes the observing apparatus.[38] If we also assume that all quanta at some point in the history of the cosmos have interacted with other quanta in the manner disclosed by the experiments testing Bell's theorem, then "wholeness" as we are defining it here is a property of the entire universe.

This situation seems "strange," as all our experience with the quantum world seems strange, in terms of macro-level expectations. Non-locality indicates that points A and B where the left and right polarizers are located in the Aspect experiments remain correlated in the unified system and, as d'Espagnat has argued, that all quanta in the vast cosmos are similarly correlated. Yet we can no more explain this scientific fact in the classical sense, or in terms of macro-level visualizations, than we can explain the quantum of action in these terms. Non-locality, like quantum transitions, is a fact of nature understandable to us only within the limits and epistemological implications of the indeterminacy principle. Our task is to say as much as we can about them based on an entirely objective analysis of efforts to coordinate experience with them. More important, we can no longer rationalize this strangeness away by presuming that it applies in principle only to the quantum world. All indications are that Bohr was correct in his assumption that we live in a quantum mechanical universe, and that classical physics represents a higher-level approximation of the dynamics of this universe. If this is so, then the epistemological situation in the quantum realm should be extended to apply to all of physics.

As we will demonstrate later, alternatives to CI are, in our view, fatally flawed in two respects—they are not subject to experimental verification and, more interesting, they involve appeals to extra-scientific or metaphysical constructs. If this is so, complementarity should be viewed as a physical principle as opposed to a philosophical position. Why physicists would elect to advance theories that violate two fundamental tenets of scientific epistemology can be largely explained in terms of an ongoing attachment to seventeenth century metaphysical dualism and the doctrine that the world is completely knowable in mathematical theory. These are the essential ingredients in what we will term the hidden ontology of classical epistemology. In order to better understand the nature of this attachment, it will be necessary to consider in more detail the manner in which the overt metaphysics of seventeenth century physics came to be infolded into the idealization that there must be a one-to-one correspondence between every element of the physical theory and the physical reality.

In the absence of that correspondence, we also face, as Bohr indicated, some large epistemological dilemmas. We are obliged to view physical theories not as an ontological bridge between observer and observed system, but rather as subjectively-based human constructs useful to the extent that they help us coordinate greater ranges of experience with physical reality. This forces us to confront once again a problem which we thought we had effectively dispensed with in positivism. If the mathematical description does not provide true, genuine, and certain knowledge in the manner suggested by Laplace and others, can physics be viewed as an entirely autonomous way of knowing? In other words, must we now include in our understanding of physical reality as it is disclosed in physics other aspects of our world-constructing minds as intervening variables in this understanding? The answer to this question, which we will explore in some detail in the next chapter, is that a commitment to metaphysical and epistemological realism does allow us, if we are particularly vigilant in keeping this commitment, to exclude linguistically-based extra-scientific variables from scientific knowledge, and to continue to view science as an autonomous way of knowing. Yet in our new situation this effort is tremendously more challenging that the positivists could have ever dreamed it could be.

5
Searching for the Ground: Mutation, Mind, and the Epistemology of Science

The very principles, in the light of which knowledge is to be criticized, are themselves found to be socially and historically conditioned.
—*Karl Mannheim*

In a quantum mechanical universe physical theories, for all the reasons we have explored, do not exist in any a priori sense outside of the mind which conceives and applies them. From this new perspective we can now appreciate the fact that the architects of classical physics were translating into mathematical language their visualizable macro-level experience with physical phenomena. As long as the mathematical description seemed completely in accord with the visualized phenomena, meaning that a one-to-one correspondence between every element in the physical theory and the physical reality appeared self-evident in experiments testing valid theories, there was no suggestion that extra-scientific assumptions were required or assumed in these theories. It was only after classical physics and its descriptive categories were extended into the realm of the unvisualizable in modern physics that the veil was lifted, and we could begin to contemplate nature in the absence of what Einstein called the "spectacles of the old-established conceptual interpretation."

Bohr, as we have seen, well understood that classical descriptive categories, like position and momentum, and energy-momentum conservation, are "distorting spectacles" which accrued as a result of our experience as macro-level observers. We must now view these categories, he suggests, as "abstractions," or "idealizations," representing "properties that are definable or observable only in interaction with other systems." Since it is both aspects of these complementary constructs which describe the complete situation, and since both cannot be simultaneously applied in a single situation, how does one fully demonstrate the validity of the categories in empirical terms? Bohr's answer is, in effect, that we cannot. The best and only thing we can do is to recognize that the classical categories are "adequate" for the "unambiguous" communication of results in modern physics. This is so, he says, because our customary points of view and forms of perception evolved as a result of experience at the macro level, and will forever be conditioned and constrained by that experience.

What this means, simply put, is that all of our conceptions of the nature of reality, including those that we derive and prove scientifically, are subjectively based. Although the pursuit of scientific truths is considerably more constrained and precise than the pursuit of non-scientific truths, the ultimate foundation of all human truths must now be viewed as identical—all truths reside, ultimately, in our world-constructing minds. Yet recognizing this, as Bohr makes clear, does not obviate the objectivity of science, its epistemological authority, or its status as an autonomous way of knowing. It merely makes the business of being objective a good deal more arduous than we previously imagined it to be.

It is no longer sufficient to merely exercise the utmost care and precision in the conduct of scientific experiments, and in the analysis of the conditions and results of these experiments. Objectivity also demands that we understand all that we can about the "prior" conditions of the experiment that exist in the "mind" of the observer in order that distorting macro-level spectacles can be either removed or more objectively understood. One way in which to accomplish this is to develop an active awareness of the prior conditions that led to the evolution of enough excess neuronal capacity in the evolving brains of our prehistoric ancestors to "invent" the symbol system of ordinary language. With some minimal understanding of these dynamics, it is easier, as we hope to demonstrate, to more effectively isolate constructs framed in ordinary language which function as distorting spectacles in modern physical theories.

Our primary motive, however, for including a brief history of the origins of human consciousness is that it serves as background for an hypothesis advanced later in this discussion having to do with the role which complementary constructs might play in "all" symbolic constructions of human reality, including those based on ordinary language. Bohr, as we shall see, intuited that complementarity was a fundamental logical principle in constructions of reality based on the symbol system of ordinary language. But he was not able, based on the relevant information available during his lifetime, to make this case as convincingly as we will hopefully be able to make it. A secondary motive has to do with our ambition to bridge the gap between C.P. Snow's two cultures with particular emphasis in this case upon some misconceptions about the character of scientific knowledge that seem to be rather widely held by social scientists and to a lesser extent by humanists.

Much of the material in this chapter is normally appealed to by those who practice a discipline known as the "sociology of knowledge." The origins of this discipline can be traced to the work of German intellectuals like Marx, Nietzsche, and Max Scheller, and to that of French intellectuals like Emile Durkheim and Marcel Mauss. The fundamental precepts of the discipline are most easily traced, however, to the work of Karl Mannheim. His essential argument is that knowledge is necessarily formed within the context of particular historical and social situations, and is, therefore, shaped primarily by those contexts. Social relations in Mannheim's view fundamentally influence the very "form" of thought, and thus epistemology is the "product" of these relations and varies from one epoch to the next.[1] Another figure whose work has been foundational in the development of the sociology of knowledge is Ludwig Wittgenstein. The two aspects of Wittgenstein's philosophy that are most consistently appealed to by students of the sociology of knowledge are that the essential logic of any sentence does not parallel or reflect the logic of reality itself, and that ordinary language with all its embedded cultural variables is the primary "determinant" of what we "regard" as real.[2] The best known figures in the sociology of knowledge among American students of this discipline appear to be Peter Berger and Thomas Luckmann. These scholars make the case that the world as it is known to us is a "socially constructed" reality that is primarily linguistically based. Although our socially constructed reality was, they argue, essentially arbitrary in its origins and tends to be rather wayward in its transformations, it legislates over and serves to legitimate virtually all aspects of human reality, including our "ideals of order" and our cosmologies.[3]

Most social scientists and humanists in our acquaintance are familiar with literature in the sociology of knowledge, and many have appealed to this literature to make the case that all knowledge, including the scientific, is highly relativistic from an epistemological point of view. This view often translates into the following conclusions: 1) scientific knowledge is itself essentially a cultural product; 2) the epistemological authority of scientific truths is not greater or lesser than that of other arbitrarily derived cultural truths; and 3) the knowledge we call scientific is primarily governed and directed by its social context. If one accepts these con-clusions, then that which passes as scientific knowledge is merely knowledge that the scientific community sanctions in its journals, reviews, textbooks, scholarly meetings, and so on. Two recent texts which seek to develop this thesis are David Bloor's *Knowledge and Social Imagery* and Bruno Latour's and Steve Woolgar's *Laboratory Life*.[4]

Part of what we will attempt to demonstrate in this chapter is that advocates of the sociology of knowledge are essentially correct in their understanding of the origins and character of the reality of everyday life which does, in fact, consist primarily of linguistically-based constructs. They are also correct, as we have already conceded and will continue to demonstrate, in their assumption that linguistically-based and socially derived constructs, like ideals of order with their usual metaphysical underpinnings, do play a role in the evolution of a scientific world-view. Yet they are clearly not correct, as we tried to suggest in the very beginning of this discussion, in claiming that science is not an epistemologically privileged way of knowing and that the alleged progress of science is driven primarily by extra-scientific, linguistically-based constructs.

Evolution of the Human Species

The history of the evolution of human consciousness should properly begin at the beginning, or with the origins of the cosmos. All of modern physics, including cosmology, implies interconnectedness and interpenetration of physical reality in space-time. Categorical divisions between stages in cosmic history have become increasingly more arbitrary. Material reality as we now understand it is most basically a manifestation of the equivalence of mass and energy in accordance with Einstein's relation $E = mc^2$. The second law of thermodynamics, also known as the law of increase of entropy, states that the universe is running down. All physical processes tend to increase entropy, or to move toward higher degrees of disorder, and all forms of energy tend to transform themselves to low grade and basically useless energy, namely, heat. Although energy is conserved overall, high grade energy is turned into low grade heat energy as the universe evolves.

As the universe expands it grows colder because this heat energy is spread over increasingly larger dimensions of space-time. Life is in part the ability to metabolize more energy than is given off, and the high grade energy wind in the form of light quanta from the sun has been its principal resource. Living organisms also take in high grade energy in the form of food, and convert it via complex biochemical processes into low grade energy. Another important aspect of all living organisms is that they can be characterized as very non-equilibrium processes which have lower entropy at the expense of higher entropy for the environment.

The first complex molecular structure capable of burning fuel biochemically, of excreting what it could not efficiently use, and of reproducing itself appears in the geological fossil records about 4 billion years ago. Here we find evidence of the existence of the ancestor of the first DNA molecule. Darwin had previously speculated that life evolved from a single source through increasingly elaborate adaptations capable of perpetuating themselves in their offspring. But it was only after modern physics allowed us to better understand the character of that source, DNA, and the manner in which mutations occur in the twisted double helix of the master molecule that evolution was put on a firm scientific foundation. The rungs in the ladder of DNA, called nucleotides, are configured in four different molecules constituting the four letters, or keys, in the genetic code. Different combinations of these rungs make up the genes, different genes constitute different instructions, and, finally, different nucleic acid instructions allowed, over time, for the formation of staggering numbers of different organisms.

Mutations are random changes in the way the nucleotides are put together resulting primarily from the chance bombardment of the master molecule by ultraviolet light from the sun, cosmic rays, nearby chemical reactions, and random quantum processes during reproduction. Most random mutations, like random behavior or accidents in daily life, are not useful, and the mutated organism will not live to perpetuate the change. But as the biologist Steven Jay Gould points out, "Evolution is a mixture of chance and necessity—chance at the level of variation, necessity in the working of selection."[5] The analogy is not that of a monkey randomly throwing bricks which over large spans of time happen to make a cathedral. The limiting condition which governs if random pulsations in the dance of life are successful is the ability of a mutated organism to produce offspring in its ecological niche. It does so in the face of incessant environmental change due in part to the effects of the success of competing organisms, and fluctuations in temperature and ecology which can also be introduced by successful adaptations. As the biologist Ernst Mayer notes, "Nothing in biology is less predictable than the future course of evolution."[6]

In understanding our own evolutionary history the intervening variables that constitute the largest single factors contributing to evolutionary success are language and culture. The rapid increase of the size of the hominid brain relative to body weight, or in the so-called encephalization quotient, was the fortunate pre-adaptive condition that allowed language and culture to evolve. But brain size is a functional adaptation to environmental conditions, and not a quantity with an inherent tendency to increase. Many big brain creatures now lie quietly in the evolutionary graveyard, and some of the more recent examples were remarkably similar to ourselves.

Our large brain is an evolutionary extension of a pattern that began with small, tree-dwelling mammals similar to rats and shrews some 75 million years ago. We shared our last common ancestor with gorillas and chimps about 6 million years ago at most. For most of the 150 million years of mammalian life the basic mammalian brain did not expand—environmental conditions that made an enlarged brain a more efficient way of perpetuating life forms with more brain power apparently arose during the period in which the dinosaurs became extinct 65 million years before the present. Moving from the period 6 million years ago when we shared our last common ancestor with the African great apes to 3 million years before the present, we discover three known species of hominids—*Australopithecus robustus*, *Australopithecus africanus*, and *Homo habilis*. Although *Homo habilis*, our ancestor, had a larger encephalization quotient and more efficient bipedalism than his australopithecine cousins, the fossil records suggest that in terms of numbers the cousins were for a time more of an evolutionary success. But *Homo habilis* was apparently possessed of an alert evasiveness that kept him on the edge of the adaptive fray, and random mutations continued to introduce useful, largely pre-patterned, behaviors related to increase in brain size.

Origin of Language

At some point the larger brain allowed our ancestors to become increasingly freed of instinctual behavioral pathways resident in their complex genetic code and to invent or learn new patterns of behavior within a "social" group. The enlarged brain also allowed these ancestors to develop roughly fabricated tools to perform a small repertoire of tasks and, most important, to invent another tool which as a mechanism of survival would make their descendants an evolutionary success par excellence. The continued refinement of this tool would dramatically alter the terms of survival, and establish a relationship between stimuli and response qualitatively different from that of any other organisms.

Language as we know it is a complicated system of signs and symbols held together by the logical structure of syntax and grammar. Its singular power is that it allows us to construct a symbolic reality based on preadaptive growth in brain size. It is also important to note that the evidence suggests that brain size was further increased by random mutations that enhanced the use of language as a survival tool. Current speculation is that the evolution of language began when the brain size of *Homo habilis* grew to roughly 600–800 cm^3. But the reality of *Homo habilis* was certainly not symbolic, and the same can be safely said of his descendent—*Homo erectus*. Although we clearly cannot reconstruct the cognitive processes in these prehistoric forebears, it is a safe assumption that language in the initial stage of development was a "sign" as opposed to a "symbol" system.

In that stage sounds produced by air from the lungs vibrating in the larynx and refined by tongue and lip movements had a one-to-one correspondence with things or events in the external environment—the sound was a sign. Our own mental landscape is constructed by and filtered through multiple layers of subtle and interrelated cognitive processes built on language and perpetuated by culture. There is in our case, as Richard Gregory puts it, no "innocent eye" because even the most seemingly "unconscious" perceptions are "conceptual" rather than "sensual."[7] The "eye" of *Homo habilis* and, to a somewhat lesser extent *Homo erectus*, was probably innocent in that sense.

Although the evolution of *erectus* was rapid in comparison with most other organisms, it was glacially slow in comparison with the rate at which *Homo sapiens* evolved from the paleolithic era into *homo sapiens sapiens*, some 100,000 years. The enlarged brain of *homo sapiens sapiens*, coupled with his more sophisticated language system, appears to have manifested its power largely in terms of more elaborate social organizations. Just as we cannot even begin to assess the complexities of social organization in hunter-gatherer and agricultural societies today in terms of their crude lithic, wood and bone implements, so we cannot assess the social organization of prehistoric *homo sapiens sapiens* on that basis. But that organization was clearly based on a growing capacity for language which allowed daily existence to be increasingly choreographed in terms of rules of kinship, ranks, roles, taboos, and obligations.

About 40,000 years ago we witness a dramatic change in the material culture of the species which has now assumed the form of modern humans—*Homo sapiens sapiens*. Representational art appears in the form of clay and stone sculpture along with simple but strikingly beautiful cave paintings. There is also a suggestion some 32,000 years ago in Europe of a form of literacy consisting of scratches on ornaments, pieces of bone and clay, and stones. Fossil records indicate that the beginnings of the artistic phase correlated with an unusual expansion of human populations in the northern hemisphere. This suggests that the ability to better coordinate experience with language and culture was proving a clearer evolutionary advantage.

Language as the basis for symbolic thinking, and the deepening of long-term memory to store vast amounts of information was, then, the penultimate product of sapientiation. Many modern linguists are convinced that language in the full sense—an arbitrary system of sound symbols that have meaning for a particular linguistic community—emerged only some 50,000 years ago.[8] This would serve to explain why art and the rapid evolution of materials-based culture appear at the same time. What is certain is that the pre-adaptive condition that allows us to assimilate language and culture is the morphological structure of our brains. Based upon studies of the relationship between increased brain size in homonids and the emergence of language and culture, anatomy professor Phillip Tobias arrived at the following formulae: "increase in brain size = gain in increase of neuronal organization = rise in complexity of nervous function = even more diversified and complicated behavior responses = progressively amplified and enhanced cultural manifestations."[9]

Evolution of Culture

The reality that is of our own making is imaged in every thought, attitude, behavior, and artifact that we witness in culture. Early cultural artifacts which had no strictly survivalistic value, like cave paintings, reach their apex 18,000 years ago. At the end of the ice age, 10,000 years ago, the human form becomes more common in these paintings, and we witness a corresponding increase in reliance upon geometric forms. Since both were rare in previous paintings, there appears to be a correlation between increased language ability and higher levels of abstract thought. That correlation would later become abundantly clear after man developed the language of mathematical physics.

The dramatic global shift in human subsistence patterns from nomadic hunting and gathering to sedentary food production at the end of the last ice age suggests that the ability to organize experience through language and culture was already fairly advanced. The presence of trade items and status burials in hunter and gathering societies during the last ice age indicates that human society was in the process of becoming more complex and hierarchical prior to the period when the earth's climate would make those skills more valuable.[10] The agricultural revolution apparently began in that arc of fertile land running through present-day Israel, Jordan, Syria, Turkey, and Iran 10,000 years ago, and another similar center appears in China 7,000 years ago.

Symbolic language, which in the initial stages freed man from the grip of pre-adaptive instinctual responses and allowed him to better coordinate his activities in the interest of survival, eventually made him free in a much more radical sense— it allowed him to create and re-create his own reality which was externalized into and perpetuated by an increasingly elaborate material and immaterial culture. These externalized constructs, like language itself, were generated in the uniquely human cerebral cortex which became increasingly more capable of exerting control over the pre-programmed and ancient emotional responses resident in the limbic system. Obviously this new reality resident in the cerebral cortex had survival value in terms of passing on human DNA to progeny. But the bulk of our creations can be classed as non-survivalistic in that they do not appear to be a consequence of the struggle for survival on the subsistence level.

In order to better appreciate the role of culture in the formation of human identity and reality, let us try to separate out the pre-adaptive conditions for cultural inheritance from the actual influences of culture. If the neuronal organization of the human brain predetermines any aspect of the shape or content of human reality transculturally, studies of human behavior in the social sciences provide few indications that this is the case. This does not mean that culture replaces genetic inheritance or that genetically controlled predispositions play no role in human behavior. Genetic research has produced substantial evidence of hereditary variation in color vision, hearing acuity, odor and taste discrimination, number and spatial ability, memory, perceptual skill, and timing of major stages in intellectual development and language acquisition. There is also some less substantial evidence which suggests that hereditary variation predisposes individuals to phobias, alcoholism, homosexuality, and certain forms of neurosis and psychosis. Yet there is no evidence that genetic inheritance provides a template for a specific range of human behaviors, or that it predetermines the character of any aspect of human reality.

A human child during the first six weeks of life cries with an immobile tongue and a fixed supralaryngeal area at the back of the pharynx. What is innate in a normal human infant is the capacity to acquire language, including a tendency to space phonemes at specific intervals, to prefer the taste of sugar, and to be visually attracted to eyes, a smiling face, and bull's-eye designs.[11] Toward the end of the sixth week the tongue is freed and the child begins to utter a range of distinct sounds. At the end of three months the babblings have no special pattern. But at six months the child begins to arrange certain sounds in sequence, accentuating some and repressing others as he patterns his responses after auditory stimuli in his environment. A child born deaf will show a gradual atrophying of the babbling behavior due to lack of auditory reinforcement. The developmental programming makes it possible at about age one for the child to utter holophrastic, or one-word, sentences, and to put words into full sentences at about eighteen months. From this point onward the inheritance of language and culture is absorbed or learned, and the raw stuff of *Homo sapiens sapiens* becomes the repository of the imaginative acts and experience of forebears who helped to fashion his reality.

The point is that the raw uncultivated stuff of our evolutionary history does not in itself produce an organism suited for survival or one that is even distinctly human. The qualities which we recognize as distinctly human are all functions of human consciousness which begins to develop fully only after children have passed through the initial stages of linguistic development. At this point an inheritance built upon but quite different from innate genetic determinants and predispositions takes over, and the emergent human being enters a reality bounded but not dictated by its large neurophysiological capacities. As the anthropologist Clifford Geertz describes the process:

> Culture, rather than being added on to a virtually finished animal, was a central ingredient in the production of that animal itself.... The perfection of tools, the adoption of organized hunting and gathering practices, the beginnings of true family organization, the discovery of fire, and most critically, the increasing reliance upon systems of significant symbols (language, art, myth, ritual) for orientation, communication, and self-control all created a new environment.... By submitting himself to governance by symbolically mediated programs, ... man determined, if unwittingly, his own biological destiny. He literally created himself.[12]

Language and Culture

Although a human child does not choose its native language nor its culture, what we call self, personality, and reality are predominantly functions of exposure to both. As modern linguists have demonstrated, the deep structure or grammar and syntax of a language system greatly determine in themselves the modes of our observations and interpretations. Rules of logic and other criteria of valid statements are greatly conditioned by the form of language and are not universal. As the philosopher Wittgenstein rightly observed, "if we spoke a different language, we would perceive a somewhat different world."[13]

The world as we learn to perceive it in ordinary language is filtered through a complex network of conceptual lenses consisting of ideas invented by countless numbers of cultural forebears. As we assimilate these ideas in childhood and adolescence, there is no suggestion that they were arbitrary in origin, or that people in the process of being inducted into radically different cultures are exposed to sets of ideas resulting in assumptions, attitudes, behaviors, and perceptions very unlike our own. Human reality as it is learned in any cultural context has the status of the "paramount" reality. And it is doubtful that any of us could go through the socialization process without experiencing severe emotional trauma in the absence of this assumption. Studies in anthropology have made it quite clear, however, that what is real for human beings is almost infinitely malleable, or that the raw stuff of *Homo sapiens sapiens*, when manipulated by the dynamics of language and culture, can become the repository of a dazzling variety of distinct reality orientations.

One of the "tendencies to occur" in the development of all languages and cultures that is particularly relevant to this discussion was witnessed in the cave paintings of the ice age. As language and culture became increasingly more elaborate, the form or meaning of the human in mytho-religious terms becomes more central and important. This emphasis tends to correspond with increased reliance upon geometric form in artifacts to image or code the character of the relationship between self and world. Those related developments in the West eventually resulted in the creation of the language of mathematical physics. And it is this early association between the new language and aspects of our mytho-religious heritage that carries over in disguised form into the metaphysical assumptions of classical physics.

Science and Religion

Religion, like science, deals in ultimacies with the difference being that the knowledge claims of science are presumably open to experiential validation or refutation in the laboratory. Also, the main emphasis of religion is upon the human being in its relation to some higher or more powerful Being or beings rather than to external physical reality. The history of science indicates that scientific knowledge and method did not spring full blown from the minds of the ancient Greeks any more than language and culture emerged fully formed in the minds of *Homo erectus* or *sapiens*. Scientific knowledge is an extension of ordinary language into greater levels of abstraction and precision through reliance upon geometric and numerical relationships.

We speculate that the seeds of the scientific imagination were planted in ancient Greece, as opposed to Chinese or Babylonian culture, partly because the social, political, and economic climate in Greece was more open to the pursuit of knowledge with marginal cultural utility. Another important factor was that the special character of Homeric religion allowed the Greeks to invent a conceptual framework that would prove useful in future scientific investigation. But it was only after this inheritance from Greek philosophy was wedded to some essential features of Judeo-Christian beliefs about the origin of the cosmos that the paradigm for classical physics emerged. Ideas from both traditions eventually flowed together in the minds of figures like Copernicus, Galileo, Kepler, and Newton. The result was a view of the relationship between physical theory and material reality which, until quite recently, successfully assumed the guise of ultimate truth.

The Hebrews, condemned it seemed to ceaseless migrations in a hostile environment, legitimated, like the early Egyptians and Mesopotamians, aspects of their evolving social order with religious cosmology. Patriarchy and the primacy of law, both of which were outgrowths of maintaining tribal unity, were deified into God the father and God the law giver. Since the children of the Father were presumed to partake of his nature or to participate in some sense in his mind, natural events, no matter how mysterious, were thought to have both cause and plan which could theoretically be explained in ordinary language. Nature to the Hebrews became a transcript of the willful and directed purpose of Jehovah, or a vast metaphor concealing omnipresent design.

In Homeric heroic religion, the gods, although presumed to have existence outside the material world, were thought to express themselves more directly in natural events. As Walter Otto puts it, the divine in this religious tradition "is not superimposed by a sovereign power over natural events; it is revealed in the forms of the natural, as their very essence and being. For other peoples miracles take place; but a greater miracle takes place in the spirit of the Greek, for he is capable of so regarding the objects of daily experience that they can display the awesome lineaments of the divine without losing a whit of their natural reality."[14] This sense of naturalism in Homer, in which the gods were identified with the processes of nature, was one of the unlikely conceptual seeds that grew into classical physics.

Greek Philosophical Thought

The Greek philosophers we now credit as the originators of scientific thought were mystics who perceived their world as replete with spiritual agencies and forces with an intensity that would be quite alien to us. The Greek religious heritage made it possible for these thinkers to attempt to coordinate diverse physical events within a framework of immaterial and unifying ideas. The fundamental assumption that there is a pervasive, underlying substance out of which everything emerges and into which everything returns is attributed to Thales of Miletos. [Theoretical physicists seeking to explain the origins of the cosmos in quantum field theory operate on the same premise but for quite different reasons.] Thales was apparently led to this conclusion out of the belief that the world was full of gods, and his unifying substance, water, was similarly charged with spiritual presence. Religion in this instance served the interests of science because it allowed the Greek philosophers to view "essences" underlying and unifying physical reality as if they were "substances."

The philosophical debate which led to conclusions useful to the architects of classical physics can be briefly summarized as follows: Thales' fellow Milesian Anaximander claimed that the first substance, although indeterminate, manifested itself in a conflict of oppositions between hot or cold and moist and dry. The idea of nature as a self-regulating balance of forces was subsequently elaborated upon by Heraclitus who asserted that the fundamental substance is strife between opposites which is itself the unity of the whole. It is, said Heraclitus, the tension between opposites which keeps the whole from simply "passing away."

Parmenides of Elea argued in turn that the unifying substance was unique and static being. This led to a conclusion about the relationship between ordinary language and reality later incorporated into the view of the relationship between mathematical language and physical reality. Since thinking or naming involves, said Parmenides, the presence of "something," thought and language must be dependent upon the existence of "objects" outside the human intellect. Presuming a one-to-one correspondence between word as idea and actually existing things, Parmenides concluded that our ability to think or speak of a thing at various times implies that it exists at all times. Hence the indivisible One does not change, and all perceived change is an illusion. The impossibility of change was not, of course, the aspect of this philosophy that found its way into classical dynamics or the laws of motion. What was incorporated is the notion of one-to-one correspondence between "mathematical" symbol and substance, and the idea of substance as indestructible. These assumptions emerged in roughly the form they would be used by the creators of classical physics in the thought of the atomists, Leucippus and Democritus. They reconciled the two dominant and seemingly antithetical conceptions of the fundamental character of being—Becoming (Heraclitus) and unchanging Being (Parmenides)—in a remarkably simple and direct way. Being, they said, is present in the invariable substance of the atoms which through blending and separation make up the things of a changing or becoming world.

The last remaining feature of what would become the paradigm for the first scientific revolution in the seventeenth century is attributed to Pythagoras, and it proved to be terribly seductive. Like Parmenides, Pythagoras also held that the perceived world is illusory, and that there is an exact correspondence between ideas and aspects of external reality. But Pythagoras had a different conception of the character of the idea that showed this correspondence. The truth about the fundamental character of the unified and unifying substance could be uncovered, claimed Pythagoras, through reason and contemplation, and this "truth" was mathematical in form.

Pythagoras established and was the central figure in a school of philosophy, religion, and mathematics, and apparently was viewed by his followers as semi-divine. For his followers the regular solids (symmetrical three-dimensional forms in which all sides are the same regular polygon) and whole numbers became revered essences or sacred ideas. In contrast with ordinary language, the language of mathematical and geometric forms seemed closed, precise, and pure. Providing one understood the axioms and notations, the meaning conveyed was invariant from one mind to another. The Pythagoreans felt that the language empowered the mind to leap beyond the confusion of sense experience into the realm of immutable and eternal essences. This mystical insight made Pythagoras the figure from antiquity most revered by the creators of classical physics, and it continues to work its magic on contemporary physicists struggling with the quantum mechanical description of nature.

The dualistic conception of reality as consisting of abstract, disembodied ideas existing in a domain separate from and superior to that of sensible objects and movements became the most characteristic feature of western philosophical and religious thought. The notion that the material world experienced by the senses was inferior to the immaterial world experienced by mind or spirit, articulated by Plato and reiterated by the Church fathers and schoolmen, has been widely blamed for frustrating the progress of physics to at least the time of Galileo. But in one very important respect it also made the first scientific revolution possible.

After the makers of that revolution began to transform the Hebrew God of the desert into the Divine Architect with the assistance of Greek philosophy, it was possible to suppose that the mental categories used to classify physical events (substance, essence, matter, form, and quantity) were transcriptions of the actual as opposed to metaphors for that which language cannot describe or contain. Believing that their minds, when discoursing in mathematical and geometric language, mirrored, however feebly, that of the supremely rational Creator of the universe, there was reason to suppose that the immaterial laws which gave form to material reality were accessible to human understanding.

The Emergence of the Classical Paradigm

Progress was made in mathematics and to a lesser extent in physics from the time of classical Greek philosophy to the seventeenth century in Europe. In Baghdad, for example, from about A.D. 750 to A.D. 1000, substantial progress was made in medicine and chemistry, and the relics of Greek science were translated into Arabic, digested and preserved. Eventually these relics re-entered Europe via the Arabic kingdoms of Spain and Sicily, and the work of figures like Aristotle and Ptolemy reached the budding universities of France, Italy, and England during the Middle Ages. For much of this period the Church provided the institutions, like the teaching orders, needed for the rehabilitation of philosophy.

But the social, political, and intellectual climate in Europe was not ripe for a revolution in scientific thought until the seventeenth century. Until well into the nineteenth century the work of the new class of intellectuals we call scientists was more avocation than vocation. The word "scientist" does not appear in English until around 1840. Copernicus would have been described by his contemporaries as an administrator, diplomat, avid student of economics and classical literature, and, most notably, as a highly honored and placed church dignitary. Although we name a revolution after him, this devoutly conservative man did not set out to create one. The placement of the sun at the center of the universe seemed right and necessary to Copernicus not as a result of making careful astronomical observations. He, in fact, made very few of them in the course of developing his theory, and then only to ascertain if his prior conclusions seemed correct. The Copernican system at the time was also not any more useful in making astronomical calculations than the accepted model, and was in ways much more difficult to implement. What, then, was his motivation and his reasons for presuming that the model was correct?

Copernicus felt that the placement of the sun at the center of the universe made sense because he viewed the sun as the symbol of the presence of a supremely intelligent and intelligible God in a man-centered world. He was apparently led to this conclusion in part because the Pythagoreans believed that "fire" exists at the center of the cosmos, and Copernicus identified this fire with the fireball of the sun. The only support that Copernicus could offer for the greater efficacy of his model is that it represented a simpler and more mathematically harmonious model of the sort that the Creator would obviously prefer. The language used by Copernicus in *The Revolution of Heavenly Orbs* illustrates the religious dimension of his scientific thought: "In the midst of all the sun reposes, unmoving. Who, indeed, in this most beautiful temple would place the light-giver in any other part than whence it can illumine all other parts?"[15]

The belief that the mind of God as Divine Architect permeates the workings of nature was the guiding principle of the scientific thought of Johannes Kepler. For this reason most modern physicists would probably feel some discomfort in reading Kepler's original manuscripts. Physics and metaphysics, astronomy and astrology, geometry and theology commingle with an intensity that might be offensive to those who expect science to be narrowly scientific in the modern sense of that word. Physical laws, wrote Kepler, "lie within the power of understanding of the human mind; God wanted us to perceive them when he created us in His image in order that we may take part in His own thoughts.... Our knowledge of numbers and quantities is the same as that of God's, at least insofar as we can understand something of it in this mortal life."[16]

Believing, like Newton after him, in the literal truth of the words of the Bible, Kepler concluded that the word of God was also transcribed in the immediacy of observable nature. Kepler's discovery that the motions of the planets around the Sun were elliptical, as opposed to perfect circles, may have made the universe seem a less perfect creation of God in ordinary language. But for Kepler the new model placed the sun, which he also viewed as the emblem of divine agency, more at the center of a mathematically harmonious universe than the Copernican system allowed. Communing with the perfect mind of God required, as Kepler put it, "knowledge of numbers and quantity."

Since Galileo did not use, or even refer to, the planetary laws of Kepler when those laws would have made his defense of the heliocentric universe more credible, his attachment to the god-like circle was probably a more deeply rooted aesthetic and religious ideal. But it was Galileo, even more than Newton, who was responsible for formulating the scientific idealism which quantum mechanics now forces us to abandon. In *Dialogue Concerning the Two Great Systems of the World*, Galileo says the following about the followers of Pythagoras: "I know perfectly well that the Pythagoreans had the highest esteem for the science of number and that Plato himself admired the human intellect and believed that it participates in divinity solely because it is able to understand the nature of numbers. And I myself am inclined to make the same judgment."[17]

This article of faith, that mathematical and geometrical ideas mirror precisely the essences of physical reality, was the basis for the first scientific revolution. Galileo's faith is illustrated by the fact that the first mathematical law of this new science, a constant describing the acceleration of bodies in free fall, was not discovered by dropping balls of different weights from a tower, like that in Pisa, and discovering that they impacted at the same time. Not only do such experiments, as Galileo admitted, fail to yield precise results, there was simply no way to subject this law to rigorous experimental proof in the seventeenth century. Galileo believed in the absolute validity of his law in the absence of experimental proof because he believed that movement could be subjected absolutely to the law of number. What Galileo asserted, as the French historian of science Alexander Koyré put it, is "that the real is in its essence, geometrical and, consequently, subject to rigorous determination and measurement."[18]

The popular image of Isaac Newton is probably still that of a supremely rational and dispassionate quester after empirical truth. Although Newton, like Einstein, had the ability to concentrate unswervingly on complex theoretical problems until they yielded a solution, what most consumed this restless intellect was not the laws of physics. In addition to believing along with Galileo that the essences of physical reality could be read in the language of mathematics, he also believed, with perhaps even greater intensity than Kepler, in the literal truths of the Bible. For Newton the language of physics and the language of biblical literature were equally valid sources of communion with the eternal and immutable truths existing in the mind of the one God. Newton's theological writings in the extant documents alone consist of over a million words in his own hand, and some of his speculations seem quite bizarre by contemporary standards. The earth, said Newton, will still be inhabited after the day of judgment, and the New Jerusalem must be large enough to accommodate both the quick and the dead. Newton put his mathematical genius to work and determined that the dimensions required were "the cube root of 12,000 furlongs."

The point is that during the first scientific revolution the marriage between mathematical idea and physical reality, or between mind and nature via mathematical theory, was viewed as a sacred union. In our more secular age the correspondence takes on the appearance of an unexamined article of faith or, to borrow a phrase from William James, "an altar to an unknown god." Heinrich Hertz, the famous nineteenth century German physicist, nicely describes what there is about the practice of physics that tends to inculcate this belief: "One cannot escape the feeling that these mathematical formulae have an independent existence and intelligence of their own, that they are wiser than we, wiser than their discoverers, that we get more out of them than was originally put into them."[19]

Hertz made this statement without having to contend with the implications of quantum mechanics. But the feeling he describes remains the most enticing and exciting aspect of physics. That "elegant" mathematical formulae provide a framework for understanding the origins and transformations of a cosmos of enormous age and dimensions is a staggering discovery for budding physicists. Professors of physics do not normally tell their students that the study of physical laws is an act of communion with the perfect mind of God, or that the laws have independent existence of the minds that discover them. But the business of becoming a physicist typically begins with the study of classical or Newtonian dynamics, and this training provides considerable covert reinforcement of the feeling which Hertz describes.

Einstein's View

Perhaps the most expedient way to examine the legacy of the western religious heritage in the metaphysics of classical physics is to briefly consider the source of Einstein's objections to quantum epistemology. The central issue for Einstein was not whether indeterminacy is an irremedial aspect of physical reality. It was the doctrine from classical epistemology that there must be a one-to-one correspondence between every element of a physical theory and the physical reality the theory describes. Einstein apparently lost faith in the God portrayed in Biblical literature in early adolescence. But as the following passage from *Autobiographical Notes* suggests, there were aspects of that heritage which carried over into his understanding of the foundations for scientific knowledge:

> Thus I came—despite the fact that I was the son of entirely irreligious (Jewish) parents—to a deep religiosity, which, however, found an abrupt end at the age of 12. Through the reading of popular scientific books I soon reached the conviction that much in the stories of the Bible could not be true. The consequence was a positively frantic (orgy) or freethinking coupled with the impression that youth is intentionally being deceived by the state through lies; it was a crushing impression. Suspicion against every kind of authority grew out of this experience.... It was clear to me that the religious paradise of youth, which was thus lost, was a first attempt to free myself from the chains of the "merely personal."... The mental grasp of this extra-personal world within the frame of the given possibilities swam as highest aim half consciously and half unconsciously before the mind's eye. [20]

It was, suggests Einstein, belief in the word of God as it is revealed in biblical literature that allowed him to dwell in a "religious paradise of youth," and to shield himself from the harsh realities of social and political life. In an effort to recover that inner sense of security which was lost after exposure to scientific knowledge, or to become free once again of the "merely personal," he committed himself to understanding the "extra-personal world within the frame of given possibilities," or, as seems obvious, to the study of physics. Although the existence of God as described in the Bible may have been in doubt, the qualities of the mind which the architects of classical physics associated with this God were not. This is clear in the following comment by Einstein on the uses of mathematics:

> Nature is the realization of the simplest conceivable mathematical ideas. I am convinced that we can discover, by means of purely mathematical constructions, those concepts and those lawful connections between them which furnish the key to the understanding of natural phenomena. Experience remains, of course, the sole criteria of physical utility of a mathematical construction. But the creative principle resides in mathematics. In a certain sense, therefore, I hold it true that pure thought can grasp reality, as the ancients dreamed.[21]

This article of faith, first articulated by Kepler, that "nature is the realization of the simplest conceivable mathematical ideas," allowed Einstein to posit the first major law of modern physics much as it allowed Galileo to posit the first major law of classical physics.

Most physics textbooks still credit the Michelson-Morley experiment, which failed to discover the existence of the ether, as something like a causal influence in the development of the special theory. But Einstein was not led to the special theory as a result of experiment any more than Galileo was led to his law by dropping objects of varying weights from the Tower of Pisa. Such explanations seem to endure because of our attachment to the false notion that the progress of science moves in lock step from experimental result to physical theory. The record indicates that Einstein was apparently unfamiliar with the results of the Michelson-Morley experiment in 1905, and would probably not have been terribly impressed even if he had known them.

As Gerald Holton has shown, Einstein's conclusion that the ether was non-existent was a consequence of his effort to eliminate "asymmetries" in Maxwell's equations on electrodynamics.[22] The desire to grasp reality through pure thought in simpler and more elegant mathematical abstractions made Einstein discomforted by the fact that there was one equation in Maxwell's theory for finding the electromagnetic force generated by a moving conductor when it goes past a stationary magnet, and another equation when the conductor is stationary and the magnet is moving. Einstein then extended the principle of relativity from mechanics to all of physics, abandoned the notion of the absolute frame of reference and, consequently, the ether. This forced him to explain why every observer regardless of his state of motion could perceive the speed of light as constant.

During the period when the special and then the general theories of relativity had not been confirmed by experiment and many established physicists viewed them as at least minor heresies, Einstein remained entirely confident in their predictions. Ilse Rosenthal-Schneider, who visited Einstein shortly after Eddington's eclipse expedition had confirmed a prediction of the general theory (1919), described Einstein's response to this news:

> When I was giving expression to my joy that the results coincided with his calculations, he said quite unmoved, "But I knew the theory is correct," and when I asked, confirmation of his prediction, he countered: "Then I would have been sorry for the dear Lord—the theory is correct."[23]

Einstein was not given to making sarcastic or sardonic comments, particularly on matters of religion. These unguarded responses testify to his profound conviction that the language of mathematics allows the human mind access to immaterial and immutable truths existing outside of the mind which conceives them. Although Einstein's belief was far more secular than Galileo's, it retains the same essential ingredients.

What was at stake in the twenty-three–year-long Bohr-Einstein debate was, as we have seen, primarily the status of an article of faith as opposed to the merits or limits of a physical theory. At the heart of this debate was the fundamental question, "What is the relationship between the mathematical forms in the human mind called physical theory and physical reality?" Einstein did not believe in a God who spoke in tongues of flame from the mountaintop in ordinary language, and could not sustain belief in the anthropomorphic God of the West. What he meant by the word God in that sense is quite ambiguous.

The doctrine that Einstein fought to preserve seemed the natural inheritance of physicists until the advent of quantum mechanics. Although the mind that constructs reality might be evolving fictions which are not necessarily true or necessary in social and political life, there was, Einstein felt, a way of knowing purged of deceptions and lies. This knowledge mitigates loneliness and inculcates a sense of order and reason in a cosmos which might appear otherwise bereft of meaning and purpose. The knowledge is physical theory which mirrors the pre-existent and immutable physical laws.

What most disturbed Einstein about quantum mechanics, as we have seen, was the fact that this physical theory does not in experiment, or even in principle, mirror precisely the structure of physical reality. There is, for all the reasons we have discussed, an inherent uncertainty in measurement of quantum mechanical processes reflected in quantum theory itself which clearly indicates that there are limits within which this mathematical theory does not allow us to predict or know the outcome of events. Einstein's fear was that if quantum mechanics were a complete theory, it would force us to recognize that this inherent uncertainty applied to all of physics. The ontological bridge between mathematical theory and physical reality would, therefore, be broken. And this would mean, as Bohr was among the first to accept, that we must profoundly revise the epistemological foundations of modern science. We see these reservations at work in Einstein's comment, "I cannot seriously believe in the quantum theory because it cannot be reconciled with the idea that physics should represent a reality in time and space, free from spooky actions at a distance."[24]

The Necessity to Use Classical Categories

Although this brief and far from adequate discussion of the origins and history of human consciousness suggests that ordinary reality can be quite mutable and varied in different linguistic and cultural contexts, it also seems to argue that the classical descriptive categories are in some sense, as some students of the sociology of knowledge have claimed, arbitrary cultural products. If the form of ordinary reality results largely from the assimilation by the creature with the enlarged forebrain of quite different languages and cultures which condition all forms of perception, and the classical descriptive categories are based on visualizable experience, then perhaps Bohr was not correct in suggesting that the categories are "adequate" for "unambiguous" communication of results in modern physics. When we also consider the extent to which the emergence of these categories was conditioned by mytho-religious heritage, Bohr's position seems weaker still.

The first argument for the adequacy of these categories, aside form the fact that they work beautifully in applying classical dynamics to the macro world, is simply that we are macro-level perceivers. It is on this level that we record results in quantum physics and on this level where we communicate them. We have the choice here of operating out of this pragmatic necessity, or ceasing to conduct experiments and to communicate results. Given this inescapable condition, the challenge becomes to use the categories in "unambiguous" communication. In the absence of the logical framework of complementarity, its rules for observations, and its explicit limits on the correspondence between physical theory and physical reality, the categories are not adequate. They become quite ambiguous. Providing, however, that we communicate them with a full awareness of all that complementarity requires and implies, these macro-level lenses become the best possible means of communication for macro-level perceivers and communicators.

But this new way of viewing the physical world also suggests that the classical categories, and indeed all of physics, are subjectively-based constructs that evolved in our efforts to coordinate experience with nature in mathematical language. Although this is, we believe, the case, this does not obviate the fact that mathematical language is demonstrably far more precise in coordinating experience with physical reality than ordinary language, and that it was the cumulative progress of increasingly more sophisticated mathematical theories that has consistently led us to an improved understanding of the actual character of physical reality. This progress obviously could not and would not have occurred if there was not a profound commitment in the community of scientists to epistemological and metaphysical realism.

It is this commitment, as we noted in the beginning, which explains in part why the progress of mathematical theories imposes increasingly tighter constraints on what can be viewed as scientific concepts, problems, or hypotheses, and also why this progress has led rather consistently in the modern period to radically new and counterintuitive results in terms of our linguistically-based and socially determined constructions of reality. Similarly, it is a commitment to epistemological and metaphysical realism in the study of the history of science that empowers us to appreciate the fact that classical epistemology featured a pre-scientific ontology which "happened" to serve the progress of science quite well. Most of the pre-scientific assumptions that were products of visualizable experience, like space and time as separate and discrete dimensions and the notion of substance as immutable and indestructible, have been undermined by modern physics. And yet the hidden metaphysical presupposition reflected in the belief in the one-to-one correspondence between every element of the physical theory and the physical reality has, as we shall see, continued to work its magic on those struggling with epistemological implications in a quantum mechanical universe.

It now appears, however, that we have reached the point in our dialogue with nature in physical theories at which a confrontation with this hidden metaphysical presupposition is quite unavoidable. Although we have been moving inexorably toward this confrontation since the 1920s, Bell's theorem and the experiments testing that theorem have definitely, for reasons we will try to demonstrate in the next chapter, brought it to a head. What is most exciting about this development for physicists and non-physicists alike is that the resolution of this conflict is likely, in our view, to result in a profound new conception of the relationship between part and whole in physical theory. This new relationship "implies" without, for reasons we will explore later, being able to "prove," that human consciousness participates in the life of the cosmos in ways that classical physics completely disallowed. If our hypothesis is correct, our attempts to coordinate experience with reality in terms of linguistically-based social constructions of this reality may be more intimately connected to the life of the cosmos as quantum physics invites us to "envision" it than most students of the sociology of knowledge could even begin to imagine.

6
Science and the Quest for a New Metaphysics

... May God us keep
From single Vision and Newton's sleep!
—*William Blake*

Although we have, as Blake devoutly wished, awakened from the depressingly mechanistic vision of classical physics, the new vision of the cosmos, based on the testimony of a number of well-known physicists, appears anything but comforting. "The more the universe seems comprehensible," writes Steven Weinberg, "the more it seems pointless."[1] "Man," says Jacques Monod, "lives on the boundary of an alien world. A world that is deaf to his music, just as indifferent to his sufferings or his crimes."[2] "Life," comments Gerald Feinberg, appears simply as a "disease of matter."[3] The sense that modern physics has presented us with a view of nature that is bereft of human meaning or purpose seems remarkably pervasive in the community of physicists. Although it is possible to make the argument that science need not have any relationship to our larger sense of meaning and purpose in the cosmos, the dynamics of human consciousness are such that we can never completely obviate the connection. We cannot escape the fact, as Schrödinger put it, "that all science is bound up with human culture in general, and that scientific findings, even those which at the moment appear the most advanced and esoteric to grasp, are meaningless outside their cultural context."[4]

Since quantum physics grandly extended our ability to coordinate experience with nature in mathematical physics, and since pragmatism has presumably divorced metaphysics from the conduct of physics, how does one account for what appears to be the metaphysical angst occasioned by quantum physics? The best explanation in our view is that metaphysical dualism has continued to play an unexamined role in the physicist's conception of nature in terms of the alleged correspondence between matter and mathematical description. It is this tacit seventeenth century presupposition that we saw at work in Einstein's struggles with the quantum measurement problem, and which carries over into the methodology of contemporary physics. In this chapter we will attempt to provide substantive validity for our earlier claim that Bell's theorem and the experiments testing that theorem have forced theoretical physicists to directly confront previously hidden metaphysical presuppositions within the normal conduct of science. A number of internationally known physicists have in our view made what we will characterize as "metaphysical leaps" in a bold attempt to resuscitate or "save" the traditional ontology of scientific epistemology, and thereby to posit an alternative to the Copenhagen Interpretation.

Holism in Modern Physical Theory

Anyone who has studied modern physics cannot escape the impression, grandly reinforced by Bell's theorem and the Aspect experiments, that the universe is a vast and seamless web of activity. Yet for many physicists the sense that the cosmos is unified does not appear to compensate for the apparent loss of a one-to-one correspondence between every element of the physical theory and the physical reality. The seemingly widespread perception of the cosmos among physicists as purposeless and meaningless is perhaps occasioned more by this loss, or the threat of this loss, than by any other implication of modern physical theories. The "vision of a transcendent union with nature," suggests Evelyn Fox Keller in *The American Journal of Physics*, "satisfies a primitive need for connection denied in another realm. As such, it mitigates against the acceptance of a more realistic, more mature, and more humble relation to the world in which boundaries between subject and object are acknowledged to be quite rigid, and in which knowledge, of any sort, is never quite total."[5]

The insurmountable problem in preserving the old ontology in the face of the overwhelming evidence disclosed by Bell's theorem and the recent experiments testing that theorem has been defined by Henry Stapp. The correlations of statistically obtained outcomes for certain angles in polarization-type experiments between space-like separated regions prove that non-locality is a fact of nature. Yet we cannot posit any causal connection between these regions in the absence of faster-than- light communication. As Stapp puts it,

> No metaphysics not involving faster-than-light propagation of influences has been proposed that can account for all of the predictions of quantum mechanics, except for the so-called many-worlds interpretation, which is objectionable on other grounds. Since quantum physicists are generally reluctant to accept the idea that there are faster-than-light influences, they are left with no metaphysics to promulgate.[6]

And this lack of metaphysics, for many physicists at least, makes the world a much lonelier place.

Assuming that light speed is the ultimate limit at which signals can travel, recognizing that any attempt to measure or observe involves us and our measuring instruments as integral parts of the experimental situation, and refusing to ontologize any aspect of physical theory, we are forced to conclude that the statistical results involving simultaneous correlations evident in the Aspect experiments can be explained only in terms of the apparent and quite strange fact that the system, which includes the experimental setup, is an unanalyzable whole. If we are concerned about what these results might mean in a more cosmic sense, we must also factor into our understanding of this situation that the universe has been evolving since the Big-bang in terms of the exchange of quanta in and between fields. Since all quanta that interact with one another in a single quantum state remain in some sense a part of a "single" system, this suggests that a universe in which quanta are entangled at every level is and remains a single quantum whole. As physicists David Bohm and Bernard d'Espagnat suggest, non-locality appears to have been an inherent property of the universe throughout its history. Holism in this new sense means that the universe on the most fundamental level is an undissectable whole, and that discreteness of objects must be, in some sense, a macro-level illusion.

The vision of the cosmos as one, or as a unified whole, has been consistently reinforced in modern physics, and the recent Bell-type experiments testing quantum predictions are certainly not the first indication that this is the case. Relativity theory suggested, for example, that matter is one form of energy in the background of the space-time continuum, and that gravity manifests itself as warped space-time. Even physicists like Planck and Einstein, who were greatly discomforted by the prospect of quantum non-locality, completely understood and embraced holism as an inescapable condition of our physical existence. According to Einstein's general relativity theory, wrote Planck, "each individual particle of the system in a certain sense, at any one time, exists simultaneously in every part of the space occupied by the system."[7] And the system, as Planck makes clear, is the entire cosmos. As Einstein put it, "physical reality must be described in terms of continuous functions in space. The material point, therefore, can hardly be conceived any more as the basic concept of the theory."[8] With the elimination of the construct of discreteness, Einstein says elsewhere, the sense of ourselves as separate from the whole is merely another macro-level illusion:

> A human being is a part of the whole, called by us the "Universe," a part limited in time and space. He experiences himself, his thoughts and feelings as something separate from the rest—a kind of optical illusion of his consciousness. This delusion is a kind of prison for us, restricting us to our personal desires and to affection for a few persons nearest to us. Our task must be to free ourselves from the prison by widening our circle of compassion to embrace all living creatures and the whole of nature in its beauty. Nobody is able to achieve this completely, but the striving for such achievement is in itself a part of the liberation and a foundation for inner security.[9]

If Einstein had lived long enough to witness the publication of Bell's theorem and the results of the Aspect experiments, there is every reason to believe that he would have accepted non-locality as a fact of nature. It is also reasonable to assume that his level of discomfort, if not anguish, in the face of the ongoing success of quantum mechanics would have risen to new heights. Although the experimental verification of non-locality is the most convincing demonstration to date of the unity of the cosmos that Einstein viewed as the "foundation of inner security," it would not have made him more secure for the reason given by Stapp. Recognizing non-locality as fact means, as Einstein would have been among the first to recognize, that indeterminacy is also a fact of nature, and that Bohr was correct in suggesting that the observer and his measuring instruments "must" be viewed as inseparable from the experimental situation. The only way in which to retain the world-view of classical physics is to presume the existence of that which cannot be proven by experimental evidence—faster-than-light communication.

The central question in this chapter is whether the one-to-one correspondence between every element of the physical theory and physical reality is possible in any theoretical framework of quantum phenomena. If it is possible, we can then presume that there is a viable alternative to the Copenhagen Interpretation, and that the mathematical description of nature, as Einstein and others have conceived it, can be sustained. What we will try to demonstrate is that attempts to preserve this correspondence not only require metaphysical leaps with consequences which in principle remain outside the realm of experimental verification. They also fail to meet the requirement that testability is required to confirm the validity of any physical theory.

The Quest for a New Ontology

According to Stapp, the three "principal ontologies that have been proposed by quantum physicists" as alternatives to the Copenhagen Interpretation, or to CI, are the "pilot wave ontology" of de Broglie and Bohm, the "many-worlds interpretation" originally proposed by Everett, Wheeler, and Graham, and the "actual event ontology."[10] Although the actual event ontology is most closely associated with Heisenberg, it proceeds along lines of argumentation suggested by Bohm and Whitehead as well. The following is a summary of Stapp's more detailed commentary on each of these ontologies.[11]

According to the pilot wave ontology, a non-relativistic universe is described in terms of the square of the absolute value of the wave function P and its phase S. The quantity P is the same as the square of the absolute value of the wave function in orthodox quantum theory and it defines the probability that the particle will be found within a given region. The phase of a wave, on the other hand, gives essential information about the way a wave should be added to another wave. This addition of waves, as we have seen, is a central feature of all wave phenomena, including quantum superposition phenomena. In the pilot wave ontology the wave function is completely defined by the two quantities P and S. The central feature of this ontology is that it leads to the existence of a "quantum potential" Q which is a mathematical function filling all space and time, or more correctly in some sense, which transcends classical space and time. Also, a velocity field is defined by the rate that the phase S changes in space. Since these two mathematical functions, P and S, are seen as sufficient to generate the individual particle trajectories, definite trajectories are retained.

What the pilot wave ontology belies, as Stapp notes, is the idea "that no reconciliation is possible, in the study of atomic phenomena, between the demands of space-time description and causality."[12] Here the physical model is completely deterministic, and all trajectories of particles are classical space-time trajectories. In a classical sense, the fundamental reality of physical constructs is retained, i.e., the mathematical theory is presumed to correspond with all aspects of this reality. Here each particle's trajectory is derived by the underlying quantum potential, and this even provides an explanation for quantum non-locality.[13] In this ontology the quantum potential is viewed as ubiquitous, or existing in all space and time, and thus the correlated results in the Aspect experiments result from interconnections within the system at a deeper level, in apparent defiance of the finite speed of light. The pilot wave ontology is related to David Bohm's ideas that the universe is an unbroken wholeness, and that the part is made "manifest" from the whole. This wholeness is, he says, equivalent to an unbroken web of cosmic interconnectedness which he terms the "implicate" or "enfolded" order.[14] The role of the quantum potential Q should be compared to Bohm's *implicate order*, which in a way also operates outside the confines of space and time. Since the quantum potential is not observable, it, in some sense, lies outside the confines of space-time although it influences the space-time trajectories of particles.

One large price that must be paid in the acceptance of the pilot wave model is that the initial conditions that must be specified to determine the quantum potential are totally arbitrary. Moreover, the model is not able to explain why some possibilities given by the wave function are realized when an observation is made and others are not. As Stapp emphasizes, this problem is "bypassed" by assuming that the model forces the particle into one branch although the other branches of the wave function are "empty," and have no seeming influence on anything physical. Although the model seeks to reconstruct the classical correspondence between physical theory and reality, it is only the probability P that is testable in the laboratory. Q, in contrast, is completely unobservable. Thus we are asked to view the quantum potential Q as having an independent existence that we cannot, even in principle, test.

In the many-worlds interpretation, the wave function is ontologized, or presumed to have an independent and unconfirmable existence, in a more radical sense. Although this theory also suffers from the rather fatal flaw of being untestable, what little appeal it has derives from its alleged economy. In this interpretation the fundamental reality in the universe "is" the wave function, and nothing else need be taken into account except for the epiphenomenal consciousness of human brains. As a measurement takes place, all possibilities described by the wave function *must be* realized for the simple reason that the wave function is "real." The empty branches in this model are not then viewed as empty—observation in one branch of the splitting universe in a brain recording one measurement is made complete by observation in another branch of the universe in a brain that registers a different reading. Accepting for the moment that fact that ontologizing the wave function in this manner takes us out of the realm of experimental physics, we discover a dilemma that if all branches are "ontologically equivalent," and the universe is a mixture of all possible conditions given by the wave equation, then how are distinct parts corresponding to distinct perceptions to be established? Put in another way, how does anything actual emerge from what is so amorphous?[15]

If we confine our discussion to only the measurement problem in quantum physics, then it is possible to obscure this dilemma. Assuming that the physical system has already separated into discrete branches, one can also assume that the element of discreteness has already been introduced into the observed system.[16] If, however, we view the wave function as a continuous superposition of macroscopic possibilties, then we should find an amorphous superperposition of a continuum of many different states. Since this translates mathematically into effectively zero probability, the existence of a conscious observer registering specific measurements in quantum mechanical experiments is quite improbable.

It is also clear that the impulse toward an economical description in this instance does not result in greater economy when we consider the vast number of parallel universes that result. If nature tends to be economical, there is precious little such economy suggested in the many-worlds interpretation. More important for our purposes, the impulse to preserve complete correspondence between physical theory and physical reality in the many-worlds interpretation obviates any opportunity to confirm that correspondence in experiment. True believing in this instance is pushed to a rather impressive extreme in that belief in the complete correspondence between physical reality and physical theory is taken as prior to, or more important than, factual demonstration of its existence.

In the actual events ontology proposed by Werner Heisenberg, which Stapp himself also favors, the fundamental process of nature is viewed as a sequence of discrete actual events. In this view it is the potentialities created by a prior event which become the potentialities for the next event. The discontinuous change of the "objective" or "absolute" wave function—which is not the same as the wave function of the orthodox view— is taken as describing the "probability" of an event which becomes an "actual" or "real" event when the measuring device acts on the physical system.[17] In other words, the discontinuous change in our knowledge at the moment of measurement is equivalent to the discontinuous change of the probability function. The real, or actual, event is represented by the quantum jump in the absolute wave function. Thus the probability amplitude of the absolute wave function corresponds with the "potentia," or the objective tendency to occur, as an actual event, and is disassociated from the actual event. The large problem presented by Heisenberg's quantum "potentia" is the failure to provide a detailed description of how this transition from possible to actual occurs in nature. And this makes the alleged quantum potentia, from the point of view of physical theory, ad hoc and arbitrary. What is ontologized in this instance is an "aspect" of the wave function, the quantum potentia, which is empowered to "choose" or "select," prior to the act of measurement, a particular macroscopic response to the measuring devices among all the available possibilities.

What is most interesting about all of these ontological models from our perspective is the decision to ontologize the wave function, or some aspect of the function. The result is that none of the models are able to present any new physical content that can be verified under experimental conditions. It seems clear that the impulse here is to sustain a metaphysical assumption by establishing a completely coherent picture of reality at the expense of direct verification in the laboratory, a seemingly contradictory task for science. These theorists have elected to ontologize the wave function in our view in order to preserve a one-to-one correspondence between every element of the physical theory and the physical reality.

It is also no accident that attempts to circumvent the alternative understanding of the relationship between physical theory and physical reality advocated by Bohr make no mention of complementary constructs. Yet if we accept the well-worn scientific precept that a physical theory is valid only if supported by the results of repeatable scientific experiments, and refuse to make metaphysical leaps, epistemological realism demands that we view the wave and particle aspects of quantum reality as complementary. Neither aspect constitutes a complete view of this reality, both are required for a complete understanding of the situation, and observer and observed system are inextricably interconnected in the act of measurement and in the analysis of results. Thus there is no one-to-one correspondence between the physical theory and the physical reality. If we elect to go beyond CI by ignoring the limitations inherent in observation and measurement occasioned by the existence of the quantum of action and seek to affirm this correspondence, we are obliged to make a metaphysical leap by ontologizing one aspect of quantum reality. This logical mistake results, as Bohr said it would, in ambiguity. And it also carries the totally unacceptable implication that metaphysics is prior to physics.

Towards a New Vision

The new epistemology of quantum theory reveals that fundamental oppositions disclosing the profound truths of nature are complementary, and those constructs have consistently brought us closer to a vision of nature which belies both classical ontological dualism and the bias that ultimate truths are transcendent and pre-existing. But the new vision of nature which discloses that the collection of processes we call "self" is not isolated from the whole also disclosed the prospect that there are no truths in the old terms. It is as if the price we paid for dispelling the notion that we are not skin-encapsulated egos was the terrifying realization that there is no empirically valid connection between our formalism for describing physical reality and reality-in-itself.

The paradox is that we are evidently one with a cosmos in which we feel existentially alone because the foundation of our being, although quite obviously a "given," cannot be completely known or revealed in the scientific description. If we demand the prospect at least of complete scientific knowledge as a prerequisite to affirming a necessary connection between our existence as intelligent beings and the life of the cosmos, then the sense of alienation so arduously and painfully communicated by the French atheistic existentialists is not merely the stuff of fiction or philosophy. There does appear to be, to borrow a phrase from Sartre's famous play, "No Exit" in epistemological terms.

It is our view that in arriving at this conclusion we are making in terms of the new epistemology of science a large, and certainly very grave, logical mistake. The ultimate extension of complementarity is that *between the whole and the part*, or the contingent and the particular. Each of these complementary constructs is necessary for the complete description of any aspect of nature, and yet one inevitably excludes the other in any particular description. What is finally different about the whole, as the results of the Aspect experiments attest, is that it has by definition no definable parts in its ultimate nature. When we consider that the validity of physical theory is dependent entirely upon measurement of the part, this conclusion is not as trite as one might initially suppose.

Science is a dialogue with nature in which we can only correlate relations between particulars, and thus any proof that the parts constitute the whole is not and cannot be the subject of science. How, then, do we even grasp the notion of a whole? The answer is, we think, quite obvious—we do so because that sense of wholeness is a "given" in consciousness. But science, again by definition, does not define wholeness any more than mathematicians can define mathematically an empty set, or cosmologists can define the universe before its origins. Definition requires opposition between at least two points of reference. Although we may, on the deepest level of awareness, sense or "feel" the underlying unity of the whole, science can only deal in correlations between the behavior of parts. It can say nothing about the actual character of the undissectable whole from which the parts are emergent phenomena. This whole is literally indescribable in the sense that any description, including those of ordinary language, divides the indivisible. For the empiricists who may feel at this point that we have suddenly taken flight into the realm of the metaphysical, after counseling against it, we should note that this conclusion will be put on firmer empirical foundations in subsequent chapters. For the moment, let us consider how it can be supported by the history of science.

The New Epistemology in a Philosophical Context

In accordance with the quantum view of the world, the analysis of any result requires the examination of all relevant prior conditions that pertain to the result. We now know, as we could not previously, that classical physics, as Ilya Prigogine puts it, was "a revealed science that seems alien to any social and historical context identifying it as the result of the activity of human society."[18] Awakening from that illusion, we discover that the process of scientific knowledge, including its origins and cultural transformations, must be included explicitly in our descriptions of physical phenomena.

With the invention of the symbol systems of mathematical and geometrical forms within the context of a culture in which ontological dualism was implicitly assumed to be an aspect of reality-in-itself, our ability to coordinate experience with nature on increasing levels of abstraction was extended enormously. The more amorphous oppositions and contrasts associated with the symbolic map space of ordinary language became oppositions between points associated with number and mathematical relations in the new language. In the seventeenth and eighteenth centuries, visualizable aspects of ordinary reality were translated into the map space of newly invented mathematical and geometrical relations—the calculus and analytical geometry. The remarkable result was that the correspondence between points in the new map space of physical theory and the actual behavior of matter in the physical landscape seemed to confirm a one-to-one correspondence between every element in the physical theory and the physical reality. It was, therefore, this correspondence that became a requirement for complete physical theories.

Since all such discoveries are, as Schrödinger suggested, "meaningless outside their cultural context," the explanation for the correspondence was a product of that context. The received logical framework for truth value in rational discourse coupled with belief in metaphysical or ontological dualism gave birth to the ontology of classical epistemology. The success of the classical paradigm coupled with the triumph of positivism in the nineteenth century served to disguise the continued reliance on seventeenth century presuppositions in the actual practice of physics. The result was, as Alexander Koyré explains, that we came to believe that the real "is, in its essence, geometrical and, consequently, subject to rigorous determination and measurement."[19] Although the reification of the mathematical idea served the progress of science quite well, it has also, as Koyré explains, done considerable violence to our larger sense of meaning and purpose:

> Yet there is something for which Newton—or better to say not Newton alone, but modern science in general—can still be made responsible: it is the splitting of our world in two. I have been saying that modern science broke down the barriers that separated the heavens from the earth, and that it united and unified the universe. And that is true. But, as I have said too, it did this by substituting the world of quality and sense perception, the world in which we live, and love, and die, another world—the world of quantity, or reified geometry, a world in which, though there is a place for everything, there is no place for man. Thus the world of science—the real world—became estranged and utterly divorced from the world of life, which science has been unable to explain—not even to explain away by calling it "subjective." True, these worlds are everyday—and even more and more—connected by praxis. Yet they are divided by an abyss. Two worlds: this means two truths. Or no truth at all. This is the tragedy of the modern mind which "solved the riddle of the universe," but only to replace it by another riddle: the riddle of itself.[20]

The tragedy of the modern mind, beautifully described by Koyré, is a direct consequence of Cartesian dualism. It is not merely an historical curiosity that Descartes claims to have arrived at his famous dualism through "revelation," as opposed to purely deductive reasoning based on empirical evidence. Descartes claimed that on December 10, 1642, he received a visit from the Angel of Truth who carried the message from God that mathematics was the conceptual key that unlocked the truths of physical reality.[21] This visit, he said, not only prompted him to invent analytical geometry, based on his studies at the Jesuit college of La Fleche, but also to formulate a new conception of metaphysical dualism. That he was enticed by the alleged visitor to invent analytical geometry is significant in that this geometry presupposes exact correspondence between number as abstraction (arithmetic and algebra), and the form of concrete physical reality in space.

More interesting, Descartes, who was quite taken with the Platonic version of metaphysical dualism in which mutability is viewed as a barrier to intelligibility, did not accept Galileo's view that empiricism was capable of proving the validity of the mathematical ideal. "In things regarding which there is no revelation," wrote Descartes, "it is by no means consistent with the character of philosophy ... to trust more to the senses, in other words to the incomprehensible judgments of childhood, than to the dictates of mature reason."[22] We discover the "certain principles of physical reality," he concluded, "not by the prejudices of the senses, but by the light of reason, and which thus possess so great evidence that we cannot doubt of their truth."[23] Since the real, or that which actually exists external to ourselves, was "only" that which could be represented in the quantitative terms of mathematics, Descartes concluded that all qualitative aspects of reality could be traced to the deceitfulness of the senses.

It was this logical sequence that led Descartes to posit the existence of two categorically different domains of existence for immaterial ideas—the "res extensa" and the "res cognitans," or the extended substance and the thinking substance. Descartes defined the res extensa as the realm within which primary mathematical and geometrical forms reside outside of our subjective reality, and res cognitans as the realm within which all other non-mathematical, and hence secondary ideas, reside in subjective reality. Given that Descartes distrusted the information from the senses to the point of doubting the perceived results of repeatable scientific experiment, how did he conclude that our knowledge of the mathematical ideas residing in the res extensa was accurate, much less the absolute truth? He did so by making a leap of faith—God constructed the world, said Descartes, in accordance with the mathematical ideas which our minds are capable of uncovering in their pristine essence.

The truths of classical physics as Descartes viewed them were quite literally "revealed" truths. His problem was then to work his way out of a conclusion which believers in validity of the experimental method would later find so oppressive in the absence of belief in the revealed character of the truths of physics—that subjectively-based non-mathematical qualities in the world where "we live, and love, and die" are purely subjective with no necessary or actual relation to reality-in-itself. It is at this point that Descartes arrives at a conclusion which from the perspective of the philosophical "implications" of quantum theory makes some sense. Direct your attention inward and divest your consciousness of all awareness of the res extensa, and the existence of the res cognitans is self-evident. We know with certainty that we are ontologically grounded in existence because the self-evident awareness of existence is prior to any definition of the scientific character of that existence.

In our ongoing dialogue with nature we have come to realize that the distinction between res extensa and res cognitans seems to be non-existent, or that the res extensa appears to be grounded in the res cognitans. Consciousness, in addition to being the prior condition for scientific thought, can also be viewed as the generative source and ground for the knowledge this thought has produced. We now know that this knowledge is, and always has been, an active construction of our conscious minds. If we are aware, as Koyré suggests, of two worlds and their disparate truths, we can no longer assume that this distinction is a given in nature. It is rather that we are constructing two versions of reality based on different criteria for the truth value of knowledge.

But this does not mean that the truth value of scientific knowledge, for all the reasons we have discussed, is diminished or compromised in the least. Although the content of consciousness in our construction of both ordinary and scientific reality is essentially arbitrary in origin, the latter is "dialogic" in a very restricted sense. In order for a scientific construct to be recognized and perpetuated as such, it must continually, as we have demonstrated throughout, stand before the court of last resort—repeatable experiments under controlled conditions. And that court, as we saw in the experiments testing Bell's inequality, will not modify its verdict based on any special pleading about the character of the defendant which is not completely in accord with the evidence presented. The primary source of our confusion in analyzing the results of the experiments testing Bell's inequality is that we have committed what Whitehead termed the "fallacy of misplaced concreteness," meaning that we have accepted abstract theoretical statements about concrete material results in terms of single categories and limited points of view as totally explanatory. This fallacy is particularly obvious in our dealings with non-locality.

Although results of experiments testing Bell's theorem "infer" wholeness in the sense that they show that the conditions for these experiments "imply" the existence of an unanalyzable and undissectable whole, the abstract theory that helps us to coordinate the concrete results cannot "in principle" disclose this wholeness. Since the abstract theory can deal only in complementary aspects of the complete reality disclosed in the act of measurement, that reality is not itself, in fact or in principle, disclosed. Although we can assume that the correlations are in some sense "caused" by the wave function, in accordance with the indeterminacy principle, or that the simultaneous results in space-like separated regions constitute in some curious way one point in space-time, these are macro-level visualizations. The distinction we are trying to make here is, in our view, a critical one. On the one hand, non-locality as a fact of nature clearly "infers" that wholeness is a fundamental aspect of nature. Yet, very importantly, neither theory nor experiment can "disclose" and, thereby, "prove" the existence of this indivisible and undissectable whole. ?

Uncovering and defining the whole in mathematical physics did seem realizable prior to quantum physics because classical theory was presumed to mirror exactly the concrete physical reality. An equally important and essential ingredient in the realization of that goal was also, however, the classical belief in locality, or in the essential distinctness and separability of space-like separated regions. That belief allowed one to assume that the whole could be described as the sum of its parts, and thus the ultimate extension of theory in the description of all the parts would "be" the whole. To put it differently, it was presumed that reductionism was valid, and, therefore, that one could analyze the whole into parts and deduce the nature of the whole from its parts. With the discovery of non-locality that picture is reversed—it is the whole which discloses ultimately the identity of the parts. Non-locality, which forces the assumption that the universe is at the most fundamental level an undissectable whole, also allows us to arrive at a somewhat different explanation as to why we can no longer posit a metaphysics for physics.

Martin Heidegger was among the first of modern western philosophers to confront the metaphysical dilemmas associated with the assumption that "One is all." At the heart of this dilemma, according to Heidegger, is the largely overlooked fact that western "metaphysics ... speaks continually and in various ways of Being. Metaphysics gives, and seems to confirm, the appearance that it asks and answers the question concerning Being. In fact, metaphysics never answers the question concerning the truth of Being, for it never asks the question. Metaphysics does not ask the question because it thinks of Being only by representing beings as beings. It means all beings as a whole, although it speaks of Being. It refers to Being and means being as beings."[24] What Heidegger is suggesting here is that instances of being, which he understands as parts, do not disclose the essence of Being, which he understands as the whole from which the parts take their existence. Thus the sum of the parts, or of beings, is not the whole—Being itself.

When we inquire into the nature of Being, says Heidegger, the proper question is not "What is this or that?" but rather "What is 'is'?" ("was ist das 'ist'?"). And this brings us into confrontation with the essential fact that "all being is in Being" or, more pointedly, that "being is Being."[25] Although it seems obvious that there is a Being which renders possible all being in that the universe, or that there is clearly something when there could just as well have been nothing, the word Being in Heidegger's view is little more than a name for something indispensable yet almost entirely indeterminate. But it is, nevertheless, abundantly true, he asserts, that we grasp the actual existence of Being with immediate definiteness and certainty. From this perspective Descartes cognito ergo sum, which recognized thought as the self-evident fact proving our existence, requires revision. The proper sequence is, "I am, therefore I think." This recognizes existence as the enabling condition of thought.

What the world-view of quantum physics discloses is not, as we understand it, a willful and cruel disregard for our need to define our ultimate relation to Being in physics, but rather the logical framework within which we will further coordinate our analysis of physical reality, or of instances of being. The framework does reveal an ultimate to the extent that it suggests that enlarged understanding of all instances of being through the process of becoming in physical reality tend toward theory represented in terms of complementary constructs. And therein we discover the even more powerful suggestion that the logical framework of complementarity is the ultimate basis for comprehending instances of being, or becoming, in the progress toward a better understanding the ground of Being, of reality-in-itself.

But the latter is obviously reality as a whole, which cannot be known, comprehended, or defined as the sum of its parts for the simple reason that the whole serves as the ground for the existence of the parts. If the whole did not participate in and serve as the ground for the existence of the parts, the parts would not exist. If one, therefore, seeks to disclose this whole by summing the parts, one is seeking to explain their existence in the absence of the ground for their existence. In other words, any effort to prove "That art Thou" and "Thou art That," in terms of logical oppositions in conscious thought, including complementarity, will necessarily fail. This is, as we will demonstrate in some detail later, a rather exact analogue for the relationship between the whole and parts featured in all modern physical theories. We will also make the case that complementarities appear in physical theories at the point at which we encounter seemingly insuperable boundaries between part and whole on various levels of emergent complexity in the life of the universe.

Does the special character of the whole as it has been revealed in modern physics suggest that we can no longer "prove" whether the life of the cosmos is, for example, "meaningless" or "meaningful?" Perhaps it does in the sense that any description of the whole as the sum of its parts does not comprehend or define the whole. Also terms like meaningless or meaningful are necessarily anthropocentric, and can be defined only within the context of an arbitrarily derived cultural context. Are we, therefore, driven to the conclusion that any felt sense of connection we have to the cosmos, represented by terms like wonder, beauty, awe, and amazement, is without meaning, or that science forbids us to embrace any of those feelings? The answer here, particularly after we more closely examine the character of the whole as it is disclosed in modern physical theory, is, as we see it, a rather emphatic "no." Yet it also true that those who would claim otherwise based on the same evidence cannot be refuted with an appeal to the evidence. What conclusions are drawn here is clearly a matter of "choice."

The central problem here as we understand it is that we have been enticed into "proving" the existence of Being when it can never be proven because of its inherent undivided wholeness. Being neither requires nor permits "proof." It merely is, and accepting this abundantly obvious fact can provide a "foundation," as Einstein put it, "for our inner security." What we have discovered is merely that the description of the parts cannot disclose the existence of the whole. Our present situation in physics suggests that we can as conscious beings analyze the parts ad infinitum with the clear and certain expectation that the whole, or Being, will not be disclosed. Very importantly, this unexpected limit to scientific knowledge seems to relegate ontological or metaphysical concerns to other domains of knowledge. Science has indeed become "metaphor" in the sense that this knowledge resides in our own world-constructing minds. But this is the metaphor that coordinates relation in the most precise terms that we can imagine. For this reason alone we can presume that science will continue to massively impact knowledge in all other areas of human thought. The "supreme fiction" that governs our constructions of reality in ordinary language will, in other words, increasingly be based on the story told by science. But this new story will be told in our view based on very different fundamental assumptions about the relationship between being and Being, and the truth value of knowledge itself.

Parallels with Eastern Metaphysics

We would be remiss in a discussion of the role that metaphysics seems to play in modern physics if we did not say something about the parallels that have been drawn between the "holistic" vision of physical reality featured in modern physics and religious traditions featuring holism, or ontolgical monism, like Hinduism, Taoism, and Buddhism. The extent to which the study of modern physical theories can entice one into embracing the eastern metaphysical tradition is nicely illustrated in a recent interview with David Bohm. In this interview, Bohm comments that, "Consciousness is unfolded in each individual," and meaning "is the bridge between consciousness and matter." Other assertions in the same interview, like "meaning is being," "all moments are one," and "now is eternity" would be familiar to anyone who has studied eastern metaphysics.[26]

Although eastern philosophies can be viewed on the level of personal belief or conviction as more parallel with the holistic vision of nature featured in modern physical theory, it is quite impossible in our view to conclude that eastern metaphysics legitimates modern physics, or that modern physics legitimates eastern metaphysics. The obvious reason why this is the case is that orthodox quantum theory, which remains unchallenged in its epistemological statements, disallows any ontology. Although modern physics definitely implies that the universe is holistic, this physics can say nothing about the actual character of the whole itself.

If the universe were, for example, completely described by the wave function, this need not be the case. One could then conclude that the ultimate character of Being, in its physical analogue at least, had been "revealed" in the wave function. We could then assume that any sense we have of profound unity with the cosmos, or any sense we have of mystical oneness with the cosmos, has a direct analogue in physical reality. In other words, this experience of unity with the cosmos could be presumed to correlate with the action of the deterministic wave function which determines not only the locations of quanta in our brain but also the direction in which they are moving. From this perspective the Aspect experiments could be providing a kind of scientific proof for ontological monism. But the problem in quantum theory is that the wave function only provides "clues" as to possibilities of events and not definite predictions of events. If, on the other hand, we assume that the sense of unity is associated with some integral or integrated property of the wave function of the brain and that of the entire universe, we confront one of the problems associated with the many-worlds interpretation. There is simply no way in the physical theory for a discrete experience of unity to emerge from the incoherently added wave functions of the multitude of parts that constitute a human brain.

Also, let us also not forget that it was the arbitrarily derived western metaphysical tradition that happened to lead to the development of the knowledge called physics. If the pursuit of that knowledge has led us to a vision which discloses more parallels with the eastern metaphysical tradition, it is difficult or, rather, impossible, to build a scientific case, based on the evidence, that this was a planned or pre-ordained progression. All we can say with any certainty from the perspective of science is that the "free" creations of two metaphysical traditions appear to have more metaphoric significance at different stages in the "bounded" progress of science.

We have been dealing thus far with challenges to cherished ideals and assumptions posed by quantum physics without a good deal of emphasis on its more positive aspects. The new epistemology, for reasons we will now begin to explore, should be welcomed not merely because it seems to be required in order to maintain the consistency and integrity of a scientific world-view. It provides, we believe, a new foundation upon which we can affirm our participation as conscious beings in a cosmos that no longer seems from the new perspective either "pointless" or "alien." Although this new foundation does not legitimate our existence in terms found in any mytho-relgious heritage, it suggests that human life and consciousness is anything but a "disease of matter." Perhaps even more important, it may even provide a new conceptual framework within which the community of man can develop a more common outlook in the interests of survival.

7
The Road Untraveled: Enlarging the New Logical Framework of Complementarity

The true delight is in the finding out,
rather than in the knowing.
—*Isaac Asimov*

As Leon Rosenfeld notes, Bohr "devoted a considerable amount of hard work to exploring possibilities of application of complementarity to other domains of knowledge."[1] Complementarity is useful in the pursuit of knowledge in other fields, said Bohr, because it allows us to "achieve a harmonious comprehension of ever wider aspects of our situation, recognizing that no experience is definable without a logical frame and that any apparent disharmony can be removed only by an appropriate widening of the conceptual framework."[2] Although the logical framework of complementarity is now appealed to in any number of disciplines, we have not even begun to apply it in the thorough and systematic fashion Bohr had in mind. When we confront a situation in which the logical framework applies, this, in Bohr's view, signals a confrontation with a profound truth.

In modern physics, as we will demonstrate in the next chapter, virtually every major advance in physical theories describing the structure and evolution of the universe has been accompanied by the emergence of new complementarities. What is most intriguing about this consistent correlation between new physical theories and profound new complementarities is that there is no suggestion that the theorists were, consciously or unconsciously, appealing to the logical framework of complementarity as a heuristic argument in formulating these theories. The necessity to recognize the existence of the quantum of action and wave-particle dualism in applying quantum physics is, of course, a kind of heuristic framework which could lead to complementarities. But even a very deliberate appeal to complementarity does not account for the actual presence of profound new complementarities in testable physical theories. And we could not, of course, even begin to explain in this manner why profound new complementarities emerged prior to quantum mechanics in relativity theory.

However remarkable this correlation might be, it does not in itself support Bohr's view that complementarity should be basic to the pursuit of knowledge in other fields. The most obvious objection is that we do not confront in other areas of knowledge the observational problem that exists in the quantum domain. Observations in virtually every other field of positive knowledge deal with macro-level phenomena, and quantum indeterminacy, although inevitably present, is quite small. We do not make our analysis of the macro-level observations "ambiguous" by failing to take into account quantum indeterminacy and the role of the observer.

Bohr apparently concluded that complementarity can be applied even in situations in which the observational problem faced by quantum theory is absent based on his understanding of the character of human consciousness in a quantum universe. Although his thesis was not fully developed, the implications are quite staggering. Taking Bohr's lead but venturing into new territory where he could not have gone due to the unavailability of supporting data to make this argument, the hypothesis here is that complementarity is the fundamental structuring principle in our conscious constructions of reality in both ordinary and mathematical languages. If the logical principle of complementarity is basic to both language systems, we could conclude that this is merely an extraordinary coincidence for which there is no known explanation. On the other hand, it might suggest that in our dialogue with nature in physical science we have uncovered a universal logical principle that is fundamental to the construction of all symbol systems in human thought. We do not know enough at present, particularly in neuroscience, to establish anything like a causal connection between complementarities in physical theories describing the evolution of the cosmos, and the apparent existence of complementarity as a fundamental structuring principle in the evolution of human thought. Yet there is sufficient evidence to suggest that this hypothesis is worthy of the closest scrutiny in future research on the physical substrates and internal dynamics of human thought.

Complementarity and Psychology

Using Bohr as our guide to this previously unexplored territory, we begin with his efforts to extend complementarity to what was then the relatively new field of modern psychology. It is here that he makes the case that complementarity is the most fundamental structuring principle in the conscious constructions of all human realities. In spite of the relative sophistication of contemporary psychology, there is one fundamental relation that remains quite paradoxical—that between thought and feeling. In pondering the character of that relationship, Bohr concluded that it was complementary. Although modern psychologists do not appeal to the logical framework of complementarity in commenting upon the manner in which thought displaces powerful emotions due to the dynamics of repression, rationalization, or denial, or the manner in which emotion displaces thought due to these same dynamics, one could make a rather convincing argument that complementarity could serve to help us better understand these dynamics.

In considering the dynamics of this relationship, Bohr came to the following conclusion: "The use of apparently contrasting attributes referring to equally important aspects of the human mind presents indeed a remarkable analogy to the situation in atomic physics, where elementary phenomena by definition demand elementary concepts."[3] Assuming that the analogy between a fundamental aspect of physical reality and a fundamental aspect of consciousness was appropriate, Bohr concluded that thought and feeling are complementary constructs. Although the presence of one tends to exclude the other in any single instance of conscious awareness, any understanding of the totality of the situation requires, as modern psychology attests, reliance on both complementary aspects of the total reality.

Whether the relationship between thought and feeling can be as neatly encompassed by the logical framework of complementarity as Bohr suggests is doubtful. But what is finally important here for our purposes is a conclusion Bohr draws in the next stage of his argument. Just as wave and particle aspects of reality in quantum physics disclose the existence of a deeper level of reality that is prior to both, so thought and feeling disclose, suggests Bohr, a deeper level of consciousness prior to both conscious attributes: "Indeed, the use of words like thought and feeling does not refer to a firmly connected causal chain, but to experiences which exclude each other because of the distinctions between the conscious content and the background we loosely term ourselves."[4] The "background we loosely term ourselves" appears identical to that level of awareness at which we sense our existence prior to any conscious content, or where we apprehend that we exist prior to any conscious constructs. Kant's term for this background was the "transcendental ego," and Heidegger would later refer to it as "Dasein."

It is against this background, suggests Bohr, that we construct the objective content of consciousness, and this process requires the use of constructs which are logically configured on the most primary level within the framework of complementarity. Since these complementary constructs derive their existence through opposition with one another, they do not manifest, as Bohr put it, as a "firmly connected causal chain." Complementary constructs are modalities of conscious awareness which, taken together, are complete formulations, or co-equal manifestations, of the content of consciousness. And they can, he suggests, be apprehended as unified, although never defined as such, on the deepest level of consciousness. Since complementary constructs by definition preclude one another in application to a given situation or, in this case, any given moment of conscious awareness, we experience them in the conscious content of consciousness as logically disparate and irreconcilable.

Obviously Bohr is suggesting that there is a level of consciousness on which we can directly apprehend the total reality which is represented in the objective content of consciousness as complementary constructs. But he nowhere attempts to ontologize that sense of unity by appealing directly to quantum physics. He did not do so for the obvious reason that quantum physics, according to the Copenhagen Interpretation, cannot directly support any ontology in that it is utterly committed, in Bohr's formulation of it at least, to metaphysical and epistemological realism. There is, however, an inference here, which Bohr apparently chose not to make explicit, that the sense of unity apprehended on the deepest levels of consciousness might have some relationship to the unified cosmos, or to reality-in-itself. As we suggested earlier, the new relationship between part and whole in modern physics also carries this inference. Although it cannot be directly confirmed by scientific theory or experiment, and cannot, therefore, be "proven" in scientific terms, it is, nevertheless, logically consistent with the modern scientific world-view.

In the same discussion, Bohr also suggests that the logical framework of complementarity as a fundamental dynamic of our conscious constructions of reality provides a basis for comprehending the dynamics of human choice. The physicist was, Bohr firmly believed, free to choose which classical construct is used in the analysis of results in quantum mechanical experiments, and the same applies on a much larger scale to the decision-making process in ordinary experience. The degree of freedom in determining, for example, which word or phrase to use in normal conversation is so large that the final decision is, said Bohr, "essentially equivalent to an improvisation."[5] Although Bohr does not appeal to quantum indeterminacy to legitimate this freedom, he nevertheless attributes the quality of freedom, or openness, to acts of cognition spontaneously generated against the "background" of consciousness.

Although Bohr's argument is by analogy, it is developed with considerable precision, beginning with an analysis of the impact of complementarity on understanding the observational process in empirical sciences. The unity of all sciences is based, said Bohr, on the shared commitment to an objective mode of description, or to a suitable framework for the unambiguous description of observed phenomena. In quantum physics, as we have seen, the properties of the observed system are not possessed by it independently of interactions with the observing system. A description which attributes properties to the observed system that exist "only" as a result of the interaction between the observer and that system is, in Bohr's terminology, "ambiguous." The seeming paradox, which the logical framework of complementarity resolves, is that non-interaction with the system precludes interaction, and yet both constitute a complete view of the situation.[6]

Bohr then proceeds to treat human consciousness "as if" it were a quantum mechanical system, and applies the logical framework of complementarity to understanding the manner in which we make "choices." He first suggests that the distinction between consciousness as the object of description, and consciousness as the subject of description, is analogous to his earlier distinction between the content of consciousness and the contentless "background" of consciousness. The analogy Bohr used in this instance is the multivalued functions for a complex variable in mathematical theory.[7] Why he may have chosen this analogy, as opposed to specific experimental situations in physics or psychology, will concern us in a moment.

The variable he asks us to consider is a function of an independent variable c, with c being a complex variable. As we saw in the first two chapters, a complex number can be written as the sum of a real and an imaginary number, $c = a + bi$, where i is the square root of -1. Any complex number can be represented as a point on a plane with the horizontal axis being the real axis and the vertical axis being the imaginary axis. On such a plane, the complex number can be represented by a set of two mathematical quantities—the length of the distance of the point from the origin of the axes, and the angle between that distance and the horizontal axis, ϕ. Functions of a complex variable, like the function log (c), can have an infinity of values, and all values cannot be plotted on the same plane without creating ambiguity. The German mathematician Riemann resolved this problem with the suggestion that such functions can be mapped as different branches of a single curve. In other words, each function could be represented on a different plane as a continuous curve which would represent a single-valued function connecting the different planes. As the angle ϕ changes by 2π, the function should be mapped on a different plane.

Bohr then suggests that the "subject self" is analogous to a multivalued function of a complex variable, and the "object self" is analogous to the process of mapping that function onto a single plane of objectivity. When we attempt to describe the subject self, we are, in effect, "mapping" that meaning onto the "plane" of objectivity in a manner analogous to mapping the complex point onto a plane of objectivity in order to determine the value of the complex function. In both examples, we are trying to translate the subject into the object. Yet in mapping the complex point, in the manner suggested by Riemann, we find that we can go full circle around the point, and yet that we do not return to the same value of the complex function. This is also true, suggests Bohr, in our efforts to translate the subject self into the object self. We perpetually construct out of the infinity of values resident in the subject self the objectified, or defined, object self. Although this object self is a function of and expresses the subject self, our maps, or descriptions, of the subject self do not and cannot contain or completely define that self. Thus any description of that which is apprehended as true for the object self is only one "objectified" expression of the subject self, and cannot be the entire description.[8] This means, simply put, that no matter how hard we attempt to truthfully and unambiguously communicate all that we know about self, we will never arrive at precisely the same description.

Complementarity and Human Language Systems

The intent here is not to suggest that clinical psychologists could benefit from a study of higher mathematics. We have developed this argument to illustrate that Bohr seems to have flirted with the prospect that complementarity is a fundamental dynamic in human consciousness reflected in the structure of mathematical language. If he had been able to make a convincing case that complementarity is a fundamental structuring principle in both mathematical and ordinary languages, then the suggestion that modern physics had disclosed an essential dynamic in the construction of all human realities or in human consciousness itself, might have been viewed as something more than philosophical speculation. Pursuing the same line of argument, we will try to show that Bohr appears once again to have been remarkably prescient.

Virtually all efforts in the modern period to uncover the underlying principles and structures which could serve to explain the manner in which an arbitrary systems of sound-symbols, or ordinary language, can have "meaning" for a given cultural community have viewed "binary oppositions" as the most primary or fundamental dynamic of this process. The best known of American linguists, Noam Chomsky, relied on binary oppositions to make the case that there is a "universal" grammar underlying all language systems which is resident in the human mind at birth. In positing the existence of a "universal" grammar, Chomsky was attempting to lend credence to the correspondence or referential view of language previously advanced by figures like John Locke. Such a view allows one to assume that linguistically-based constructions of reality are not, as Nietzsche proclaimed, "prison houses" within which we are locked within our own uniquely subjective and intensely private constructions of reality with no real or necessary correspondence with external reality.

Chomsky challenged the efficacy of what was then the two dominant models designed to explain linguistic regularities—finite-state grammars and phrase-structure grammars. In finite-state grammars the assumption is that sentences are formed in a linear fashion—one word leads to the selection of a second and a third and so on. Phrase-structure grammars, in contrast, were based on the assumption that words are regularly grouped with other words, or phrases, and that these phrases could be broken down into a logic of subphrases. Chomsky first demonstrated that finite-state grammars simply cannot account for the linguistic content of many sentences and that phrase-structure grammars were similarly limited. His alternative model, transformational grammar, assumes that the syntax of language, or the logical connections between types of words, could be studied independently of semantics, or of what the words actually mean, in terms of the principle of binary oppositions. In this way he was able to posit the existence of a universal grammar that could be studied in the "abstract," or divorced from individual speakers of language.[9] When traditional linguists after Chomsky confronted the apparent fact the formal analysis of grammar does not adequately explain language as it is used by actual speakers, they have typically appealed to a distinction between performance, or language as it is described by linguists, and competence, deviations from the theoretical system due to the "limitations" of individual speakers.

The competing and alternate views on the nature and function of language that have arisen in the modern period, although they continue to rely on the assumption that binary oppositions are the fundamental structuring principle in all language systems, have rather consistently moved away from the notion of a "universal" grammar. The prevailing view now seems to be that there can be no such grammar primarily because language exists only as it used by actual speakers. We also witness here rather consistent revisions of the character of binary oppositions as the fundamental structuring principle in language.

The first phase in this revision process was undertaken by Ferdinand de Saussure in the *Course on General Linguistics*,[10] where he argues that a linguistic sign is composed of a signifier, or the "sound" of the word, and the signified, or the "concept" that exists in the mind. In contrast with Locke, and in concert with the views of Wittgenstein, Saussure came to the conclusion that the structure and dynamics through which "meaning" is generated in language systems do not appear to correspond with those of external reality. As Saussure puts it, "In itself, thought is like a swirling cloud, where no shape is intrinsically determinate. No ideas are established in advance, and nothing is distinct before the introduction of the linguistic structure."[11] The work of Saussure would later become seminal in the development of linguistically-based theories about the character of human reality in a movement known as structuralism. The most significant assumption among the group of diverse scholars who came to be identified with this movement is that it is language that structures human reality, and all that we can know about this reality is "determined" by language. This same assumption was also adopted, as we saw earlier, in the discipline known as the sociology of knowledge which has been influenced in no small part by the work of the structuralists.

Thought as the structuralists conceive it does not preexist language and is necessarily contemporaneous with language. The primary process through which meaning is constructed is in their view via the opposition between the signifier and the signified which results in the production of signs. The most significant aspect of this opposition for Saussure and all the other theorists who have appealed to it is that any linguistic sign is "arbitrary" in that there is no apparent connection between the sound of the word, signifier, and the concept it conveys, signified. This means, in short, that language is non-referential, or has no necessary connection with reality as it actually is. What can be known is only language itself, or the total system of acoustical differences that structure thought.

Since Saussure's conception of the binary opposition between signifier and signified became the primary structuring principle in Roman Jakobson's studies of phonology, in Levi-Strauss' anthropology, in A.J. Greimas' studies of the structure of narratives, and in Roland Barthes' critical studies of drama, these figures came to be identified as structuralists. All of these scholars presumed, like Saussure, that binary opposition is a fundamental dynamic of the neuronal processing system of the human brain, and that it is, therefore, reflected in all structures of meaning. The view of the character of binary oppositions would later be revised by a group of theorists who came to be known as post-structuralists and/or deconstructionists— Michel Foucault, Jacques Lacan, and Jacques Derrida.

In the work of Foucault and Lacan, the static binary oppositions of Saussure and Jakobson in which signifier and signified are somehow "bounded" together is refuted with the assumption that there is an inevitable "space" or "gap" between them. We also witness here a movement away from the linguistic model for language toward a literary model with a corresponding shift from "sign" to "text" and from "language" per se to "discourse." Foucault calls into question both the possibility of any objective knowledge for human discourse and the existence of an independent subject for whom knowledge is acquired. In *The Order of Things*,[12] he makes the case that we "create" ourselves in the metaphorical space between word and object, or in the space between signified and signifier. It is this "space" or "gap," according to Foucault, within which individual subjectivity is "generated," and any one such subjectivity is radically different from another. For this reason, Foucault also argues that there is no difference between man, or mankind, as a philosophical concept and the subjectivity of one particular person as philosophical concept.

Lacan's formulation of the space or gap between signified and signifier is expressed, interestingly enough, as a kind of algebraic function, S/s, with the signifier placed over the signified. What the "bar" in this function represents, he asserts, is a "barrier resisting signification."[13] In Lacan's view not only are the inquiring subject and object of inquiry "constituted" within the language of inquiry—the signified itself (concept) is forever evasive because the only way we can attempt to get at it is via other significations. What the bar between signifier and signified represents, in other words, is a no-thing, a nothing, or a void. What Lacan has apparently sought to prove is that Nietzsche's "prison house" of language might be better imaged as a "labyrinth" within which investigations into the actual character of subject and object lead endlessly into new and different corridors with no hope of arriving at a coherent and consistent definition or understanding.

Derrida, who also accepts the view that a space or gap exists between signifier and signified, argues that we have sought to disguise this space or gap historically with what he terms "transcendental signifieds," or by appealing to notions of essence, truth, being, God, consciousness, and so on. Language, in his view, is a system of "differences" without any "positive terms," or terms which have an independent or "freestanding" meaning. Since no signified concept can be "present in and of itself," every concept is "inscribed in a chain" or within a play of "differences."[14] This means, he argues, that any sense we have of "immediacy" in thought or perception is a derived notion, or a product of a "chain of signifiers." What Derrida has essentially done here is to make binary opposition an even more dynamic concept, or, if you will, a kind of chain reaction in which the play of resemblances between oppositions is constant.

In the effort to describe this "systematic play of differences, of the spacings by which elements are related to each other,"[15] Derrida coined the French neologism "différance." "Différance" is not, he says, the name of anything. It is "literally neither a word or a concept."[16] It has "neither existence or essence. It derives from no category of being."[17] The term is used by Derrida to reinforce his thesis that every linguistic element exists only because of "what it is not" in that each element contains a "trace" of what is absent. Thus our linguistically-based constructions of reality in the present are built upon signs which have traces that connect them with linguistic elements that accrued in the past. This means, he suggests, that what we call the present can never be "immediate," or that we can never encounter the present free from the "traces" that connect us with the past. Différance, he suggests, is the fundamental dynamic of all linguistically-based constructions of reality, and, as such, "is older than the ontological difference or than the truth of Being."[18] What we have here, in short, is another argument that we are trapped within the labyrinth of language where we must endlessly wander with no hope of returning to the same place, and with no real or necessary connection with reality itself.

If these arguments seem rather arcane to the point of having little or no relevance to actual experience, we should note that they have become the foundation for what appears to be the dominant school of literary criticism in the United States and much of Europe, and are widely used by professors of literature in our colleges and universities in classroom instruction. Although the reasons why many literary critics and/or professors of literature have embraced this view are complex and deeply rooted in a number of ideological, philosophical, political, and social developments, the attempt to defend the integrity of the discipline by emulating the empirical and metaphysical realism and methods associated with the hard sciences is certainly a large part of the explanation. What makes this ironic, for reasons we are about to explain, is that if humanists had been more aware and appreciative of the truths of science, structuralism, post-structuralism and deconstructionism, if they had developed at all, would not have resulted in this terribly oppressive and nihilistic view of the character of human reality.

What is remarkable here in our view is that virtually all of the original examples used by Saussure in support of his conception of the binary opposition between signified and signifier, and virtually all of the examples that have been subsequently used to support modifications and refinements of this opposition, could have and should have been viewed as complementary constructs. Signified and signifier, or the sounds of words and concepts, are, first of all, rather obvious examples of complementary constructs. One excludes the other in a given situation or act of cognition in both operational and logical terms, and yet the entire situation can be understood only if both constructs are taken as the complete view of the situation. Although there is not enough space here to illustrate that complementarity definitely applies to the uses which the other theorists discussed here have made of Saussure's conception of the binary opposition between signifier and signified, let us provide at least a few quick examples.

Greimas in *Semantique Structurale* attempts to show that the elementary concepts in human thought are functions of binary oppositions like up/down, dark/light, and right/left.[19] Similarly, Levi-Strauss in *Le Cru and le cuit* attempts to represent fundamental binary oppositions in his studies of the deep structure of cultural organization by developing a list of such oppositions which "divide into mutually exclusive and exhaustive categories" like fire/water and Sun/Moon.[20] Without belaboring the point, these are, quite obviously, complementary constructs.

When the deconstructionists raised the objection that binary oppositions between signifier and signified are not "bonded" together as some structuralists had suggested, they were correct in the sense that one tends to exclude the other in any single act of cognition because they are complementary constructs. Since the deconstructionists did not apparently realize, however, that they were dealing with complementary constructs, they moved in another direction—they posited the existence of a space or gap between the oppositions. And it is this which leads finally to the rather disturbing notion advanced by figures like Lacan and Derrida that a close examination of the structure of linguistically-based constructions of reality leads ultimately to the realization that the constructions reduce to a no-thing, a nothing, or a void. The way out of this terrible solipsism is, in our view, rather simple and direct—all one need do is to realize that the fundamental feature or dynamic in all linguistically-based constructions of reality is, as Bohr suggested, the logical framework of complementarity.

That the logical framework of complementarity is also a fundamental feature or dynamic in the language system of mathematics is a case that is more easily made. One obvious fundamental opposition in mathematics is that between real and imaginary numbers. As we saw earlier, imaginary numbers can all theoretically be formed from the first imaginary number i, the square root of -1. But a mathematical operation in which we take the square root of a negative number does not make logical sense within the framework of real numbers. Similarly, real numbers are represented analytically as points on an infinitely extending straight line, and there is no way in which to represent real and imaginary numbers on the same line. Yet real and imaginary numbers constitute the complete description of this aspect of mathematics, and they can be represented by using higher dimensions on the complex plane.

A similar and equally fundamental complementarity exists in the relation between zero and infinity. Although the fullness of infinity is logically antithetical to the emptiness of zero, infinity can be obtained from zero with a simple mathematical operation. The division of any number by zero is infinity, while the multiplication of any number by zero is zero. What is suggested here is that as we perform these operations we are translating into the conscious content of mathematical language a primary complementarity.

A more general but equally pervasive complementarity in mathematical language is that between analytic and synthetic modes of description. Analysis, the breaking up of whole sets into distinct mathematical units, is logically antithetical to synthesis, or the bringing together of many units to form a mathematical whole. Analysis is the operative mode in differential calculus where a continuous function is divided into smaller and smaller parts resulting in the infinitely small differentials. The complementary mode in integral calculus involves the addition of infinitely small differentials to obtain a continuous function. One operation cannot be performed simultaneously with the other, and yet both constitute the complete view or analysis of a given situation.

One can also consider two branches of mathematics as disclosing a complementary relationship—geometry and algebra. Even though the two seemingly become unified in analytic geometry, the rules of geometry and algebra remain very different and appear as complementary. Drawing a geometrical figure, like circles, triangles, etc., requires the use of a ruler and a compass. Proving a geometrical theorem involves finding relationships between geometrical quantities such as angles, etc. In algebra, however, one deals with numbers rather than figures. The expression of the numbers is not important, and one does not need to "draw" the numbers. Proving an algebraic theorem involves finding relationships between numbers expressed in short-hand notation as mathematical equations. In analytic geometry one cannot, of course, employ both methodologies at the same time. A function like $y = 3x - 1$ can be represented as a geometrical figure—in this case a straight line—but then one loses the algebraic relationship between the two variables, x and y. Conversely, that relationship can be studied with no reference whatsoever to geometrical figures. For example, x and y could be real or imaginary, or themselves functions of other variables.

If the presence of the logical framework of complementarity is as fundamental to our constructions of reality in mathematical language as we have suggested, then perhaps we might discover here an answer to a large question we have confronted throughout this discussion. Why is there a correspondence between physical theory and physical reality, or between the mind capable of conceiving and applying mathematical physics and the cosmos itself? Physicists, as we have seen, are quite disturbed that we cannot answer this question in the old terms with an appeal to the metaphysical presuppositions of classical epistemology. Even the widespread acceptance of the essential unity of the cosmos disclosed in modern physics does not, in most instances, compensate for the feelings of loss associated with the demise of the classical view of the universe. Yet as long as the quantum of action is fact, there can be, for all the reasons we have explored, no one-to-one correspondence between physical theory and physical reality.

But could it be that our discovery that the quantum of action is fact has led us to a deeper, and perhaps far more satisfying, sense of correspondence between our knowledge of reality in physical theory and reality-in-itself? Although reality-in-itself is not disclosable, the suggestion is that we have been successful in coordinating greater levels of experience with that reality because the fundamental logical principle in nature is identical with that in all active constructions of human reality. Since the form of mathematical language allows for greater logical precision and internal consistency than that of ordinary language, the logical framework which lies at the base of mathematical language should extend more pristinely throughout its framework. It should follow, therefore, that the mathematical description of nature in physics should be more in accord with the actual behavior of events in nature. And so it is. It also seems clear that none of this would be the case if the quantum of action were not a fundamental reality in a quantum mechanical universe.

Complementarity and Neuroscience

If, however, we ever hope to prove the substantive validity of these claims in sufficiently empirical terms to merit the label scientific, we will obviously require something more than arguments by analogy. If complementarity is a fundamental dynamic in the construction of all human realities, and there is some linkage between the mind that constructs this reality and the dynamics of life in the cosmos itself, one obvious place to look for both is neuroscience. It is not sufficient to suggest that the human brain evolved in a quantum mechanical universe, and reflects, therefore, in its constructions of reality in language systems fundamental dynamics of the life of the universe. We need to be able to demonstrate on various levels of empirical analysis of actual brain function that this is the case. Although this is not something which the state of knowledge in neuroscience at this moment in time will allow us to do, there are some recent developments in this area which suggest that these connections could eventually be made.

The barrier of complexity in nature represented by the human brain is such that it has not yet been subjected to the level of detailed mathematical description required for scientific knowledge in the full sense. We are in the process, however, of achieving an improved knowledge of the fundamental units and processes of brain function which suggests that this knowledge, in principle at least, is attainable. All the neurons in the human brain work by receiving messages from, and sending messages to, selected recipient nerve cells. These sending cells and receiving cells are linked to one another in circuits.

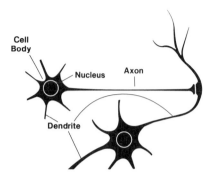

FIGURE 19. Neuron illustration showing the individual cells and their connections.

The actual link, or the place where the nerve cells communicate, is a specific point on their surfaces called the synapse. When neurons communicate through synaptic transmission, the sending cell secretes a messenger-bearing chemical onto the receptive surface of the recipient neuron. This chemical messenger, called a neurotransmitter, serves as the molecular medium for the message, and completes the circuit by carrying the message chemically across the synaptic space or gap. Neurotransmitters act on the surface of the recipient neuron by linking to a specific receptor site, in a kind of "lock and key" arrangement. The effect is either to facilitate the development of an electric current in the neighboring neuron, or to render an electric current less likely to develop. At any one moment, the chance of a neuron firing depends on the sum total of excitory and inhibitory neurotransmitters.

Estimates are that the human brain contains from 10 billion to a trillion neurons, and each appears connected to some 50,000 other neurons in a complex, three-dimensional maze. Based on the most conservative estimates, 10 billion cells with 10,000 connections each, the total number of synaptic connections in the brain would be a staggering 100 trillion. In spite of this enormous complexity, everything the brain does, from the perspective of neuroscience, is assumed to be ultimately explainable in terms of specific nerve cells, their neurotransmitters, and their responsive cells. And all the mechanisms involved are assumed to obey the laws of the physical universe, including electromagnetism, hydrodynamics, and quantum physics.

One of the most exciting of the recent discoveries about the human brain is that it works primarily via complex patterned neuronal events in large and varied brain cell populations. The old mechanistic model, which assumed that particular brain areas were dedicated to specific functions or types of information processing wired together and coordinated by some complicated linear processing network, has broken down. We now know that there are considerable overlaps and redundancies in brain function, or that the collection of parts appears in some yet to be understood way larger than the whole. For example, many regions of the brain, in addition to the cerebral cortex, are critical to learning, and memories within the cortex appear highly distributed and redundant. The cerebellum, previously thought to be involved only in the coordination and control of complex movements, is now known to store a variety of classically conditioned responses. Although there are four auditory areas and no less than twelve visual areas, overlaps and redundancies prevent us from treating these areas as dedicated components in the overall mechanisms of hearing or seeing. In this distributed system, there is no "center," and so-called brain areas are merely terms representing extrinsic connections to other features of the dynamic process.

In computer terms, the brain appears to be a massively parallel network, and it is this structure that allows for its great redundancy as well as its fault tolerance. Knowledge appears widely distributed throughout the network, and is not localized in the equivalent of a specific magnetic memory core or the position of a micro-switch. Since this knowledge seems distributed along the strengths of connections between the units, the brain does not appear to as much "compute" a solution, as we normally use that term, as to "settle into" a solution as a result of massive parallel processing.

The highly distributed character of memory and consciousness can be briefly illustrated in the example of patients who experience extensive right or left hemisphere damage, or who have had an entire hemisphere removed. If the patient is an infant, there is no noticeable impairment of memory, or any radical change in cognitive ability generally. Yet if the patient is older, say a twelve-year-old child, a wide variety of disabilities affecting speech, touch, vision, or movement, will result. The same plasticity of brain function has been demonstrated in laboratory experiments on animals. If, for example, we teach a rat to run a maze, and then remove over 50% of its cortex, it will still run the maze without significant errors. In more complex mazes, the rat will do less well, and performance will drop off in proportion to the amount of cortex removed.

This ability of the neuronal organization of the brain to act holistically is the greatest single mystery facing neuroscience. How is it, in other words, that the synaptic connections allow for the emergence of such widely distributed neuronal patterns which overlap and interweave in such a fashion as to allow the emergence of a reality, or universe of meaning, within which an actor called self wills, imagines, feels, and thinks in such a staggering variety of conceptual frames? Although the answers to this question appear to be much farther away in time than a unified field theory, many cognitive scientists and neuroscientists are convinced that quantum theory could provide a large part of the explanation. If there is a quantum mechanical wave-phase aspect on the level of synaptic connections that proves to be fundamental to a mathematical model of overall human brain function, this could provide the now missing substantive validity to the claim that com-plementarity is a fundamental aspect of human consciousness.

Henry Stapp has recently provided some small measure of this missing substan-tive validity by pointing out that the chemical transfer of a signal at the synaptic connection between neurons is apparently triggered by a small number of calcium atoms. If the number of these atoms, or other atoms that are similarly involved in this chemical transfer of information, are sufficiently small, then this suggests, according to Stapp, that quantum mechanics could provide some fundamental insights into the physical dynamics of human consciousness.[21] This thesis is also now being advanced by another physicist, Roger Penrose. If this is the case, then wave-particle dualism, complementarity, and non-locality could be instrumental in arriving at an understanding of global brain function. At the moment, however, this must remain a warranted speculation at best.

The evidence from neuroscience that best supports the hypothesis that complementarity is fundamental to brain function at present are the results of the split-brain research conducted by Roger Sperry and his colleagues on sixteen patients who had undergone a radical surgical procedure for the control of life-threatening seizures.[22] The procedure involved severing the corpus callosum, a 200-million fiber network connecting the left and right hemispheres of the brain, in the hope of confining seizure discharges to only one hemisphere. Since those results are fairly well known, we will merely summarize the most significant finding. That finding was that the two hemispheres in all normal human subjects are specialized. The left, or dominant hemisphere in most subjects, specializes in verbal, analytic, sequential thought processes, and the right specializes in non-verbal processes, such as attention, subtle pattern discrimination, line orientation, and the detection of complex auditory tones. In other words, the left hemisphere is more adept in analytical, or more narrowly logical, acts of cognition in which words are excellent tools, and the right in more holistic, or gestalt, acts of cognition in which spatial relations are excellent tools.

The adaptive advantages of this dual-processing system in evolutionary terms is that each system can specialize in different information-processing strategies which run in parallel. The dual system also allows for the independent control of limbs, like the human hands, which had obvious survival advantages for the tool-making animal. The overall advantage is that the system allows greater speed in processing information which allows for the emergence of more complex behaviors. The obvious disadvantage is that the dual-processing system is housed in a single organism which can pursue only a limited number of actions in achieving one set of goals. If the responses are to be fast, consistent, and effective, the two systems require a central coordinator. This could be achieved by allowing only one system to output functions, or by putting it in executive control. What appears to be the case is that the left, or dominant hemisphere for most of us, has overall control in most situations, and suppresses right hemisphere input. The frontal lobe appears to play an essential role in reading out the action, or behavior, that will be acted upon. It is also significant in evolutionary terms that hemispheric specialization has not been detected in the higher primates, and that it appears to have been one of the large preadaptive conditions that allowed for the invention of language.

Without belaboring the point, it is clear that complementarity applies in the effort to understand the dynamic interplay between cognitive styles in the left and right hemispheres. The two styles, for many of the same reasons that appear in the discussion of complementary relations in mathematics, displace one another in application to a given situation. Yet both constitute a complete view of the situation in the entire conscious context. If the meaning that we attempt to encode in ordinary and mathematical languages of a given situation is in some sense a function of these complementary aspects of our more complete consciousness of this situation, then it might follow that the quest for meaning, given the incessant interplay between the two complementary styles, would feature complementarity on the most basic level of oppositional constructs. Although we cannot with this argument make any direct link to complementarity in quantum physics, it does at least provide a more substantive way of explaining why complementarity appears to be the fundamental logical principle in both ordinary and mathematical languages.

Neuroscience also provides in a more approximate way some factual content in support of Bohr's distinction between the subject and the object self. The notion of self as an objective single entity capable of disclosing to itself all subjective states and/or the entire content of its own memory seems highly doubtful from the perspective of neuroscience. Any objective awareness of self is better understood in terms of a priority decision-making and action system which re-represents information from representational systems elsewhere in the brain. Our objective awareness of self appears, in other words, to be a selected subset of the full content of consciousness—it is this subset that we read out as the content of self in any given moment.

Studies of the neuronal pathways of the brain have shown, for example, that those associated with linguistic coding have limited connectivity with emotional systems. There are neural systems capable of comprehending external events and regulating organized interactions with the world that operate outside of the neural systems we associate with conscious awareness. Equally interesting, the conscious self does not appear to ignore the action of these other neural systems after it becomes aware of them—it apparently seeks to defend the evolving integrity of linguistically-based neuronal patterns associated with the unity of self by rationalizing bodily experiences from non-conscious systems. If, as Bohr suggests, our efforts to map the subject self, or the full content of consciousness, on a plane of objectivity as the object self never return to the same point, neuroscience provides some basis for understanding why that might be the case.

Complementarity and Biology

The area of science outside of quantum physics where Bohr made the most consistent effort to apply complementarity in the hope of resolving an otherwise irresolvable opposition was biology. The vitalism-mechanism controversy was a preoccupation of his father, a professor of physiology at the University of Copenhagen, and a topic frequently discussed with other academics at the family residence in the presence of Niels and his brother Harold. Although the terms Bohr used to describe that controversy may seem archaic, the distinction between a living organism, which by definition interacts with its environment, and a detailed scientific description of that organism, which must treat the system as isolated or isolatable, remains problematic. And the progressive displacement of the old and approximate biological descriptions by the more detailed and exact descriptions of modern physics has not made this problem any less problematic.

Our ability to read the code book of an organism's life and to manipulate the genetic makeup of new organisms obviously suggests that we have made great strides in improving our scientific knowledge of these processes. But no matter how precise and useful the scientific description might be, that description necessarily isolates the part from the whole in the macro domain, and thus can never be entirely unambiguous. Bohr's approach to the vitalism-mechanism duality was the same as that which he used to resolve ambiguities in quantum physics—analyze the conditions for observation required for unambiguous description, and avoid appeal to extra-scientific or metaphysical constructs.[23] He suggests that when the situation is analyzed in this way, it is difficult, if not impossible, to avoid the conclusion that we are dealing with a very fundamental complementary relationship analogous to that encountered in wave-particle dualism:

> The incessant exchange of matter which is inseparably connected with life will even imply the impossibility of regarding an organism as a well-defined system of material particles like the systems considered in any account of the ordinary and physical chemical properties of matter. In fact we are led to conceive the proper biological regularities as representing laws of nature complementary to the account of the properties of inanimate bodies....[24]

Bohr's conclusion that living organisms, characterized by their biological regularities, display an active and intimate engagement with their environment that is categorically different from that of inorganic matter is quite valid. Since organic matter and inorganic matter are constructs which cannot be applied simultaneously in the same situation, and since both are required for a complete description of the biological situation, they are, quite obviously, complementary. Not as obvious is what Bohr means by the assertion that biological regularities in organic matter represent laws of nature that are complementary to those accounting for the properties of inorganic matter. The inference is that the laws of mathematical physics can only fully describe the inanimate in that the application of these laws requires that we isolate the system in the act of making measurements. Since the biological regularities of organic matter cannot be treated as isolated, the description of organic matter in mathematical physics breaks down precisely at the event horizon at which those regularities come into existence.

A "complete" description in mathematical physics of all the mechanisms of a DNA molecule would not, for example, be a complete description of organic matter for the obvious reason that the quality of life capable of being generated via the known mechanism of DNA replication exists outside of the objectified description in the seamless web of interaction of the organism with its environment. Thus we must conclude, as Bohr apparently did, that the laws of nature accounting for biological regularities, or the behaviors we associate with life, are not merely those of mathematical physics. This problem is not obviated, incidentally, even if we are eventually capable of replicating all of the fundamental mechanisms of biological life in the laboratory by the skillful manipulation of inorganic matter. The only way in which we could assume, for reasons we will explore more fully in a moment, that there were no other physical laws involved would be to create this life in the absence of any interaction with an environment in which the life form sustains itself or interacts.

If we assume, like most physical scientists, that the mechanism of biological life can be completely explained in accordance with the laws of mathematical physics, there are a number of problems in modern evolutionary theory which in Bohr's terminology would have to be called "ambiguous." If, for example, complex biological molecules evolved from the primordial soup as a result of a purely random process, a time scale much longer than 4 billion years would be required to allow for the impressive success of biological life.[25] Similarly, the apparent compulsion of individual organisms to perpetuate their genes, in spite of recent efforts to make the case that genes are "selfish," is obviously a dynamic of biological regularities that is not apparent in an isolated system, and which cannot be described in terms of the biochemical mechanisms of DNA, or any other aspect of isolated organic matter. In the words of noted biologist Harold Morowitz, "biological systems possess history."[26] The specific evolutionary path followed by living organisms is unique, and cannot be completely described by an a priori application of the laws of physics. This is in contrast to physical macrosystems which, if prepared over and over again in identical initial conditions, will appear to evolve deterministically. One large, and quite intriguing, exception to this rule, which will concern us briefly later, is the physics of deterministic chaos.

Since what is fundamental here is the assumption that biological life cannot be treated as an isolated system, let us try to better illustrate why this is the case. Suppose, for example, that we construct an enormously elaborate computer model of all the variables that might account for the symbiosis and cooperation abundantly evident in the earth's ecosystem. Our impulse here would be to isolate the system called "life" by modeling its dynamics within the larger life system that is the ecosystem. Let us also assume that our computer model could represent all variables in the evolution of the ecosystem. Would this impossibly elaborate program allow us to fully explain the mechanisms of symbiosis and cooperation as well as competition between species?

It could not. Most obviously, the ecosystem, like any system, cannot be isolated from the rest of the cosmos in a quantum mechanical universe. Suppose, however, we seek to obviate that theoretical truth with the argument that we are dealing with macro-level processes, and thus the speed of light and the quantum of action need not concern us in arriving at practical or workable results. This argument will not save the conditions for our isolated experimental situation for two reasons. First, the indeterminacy of quantum mechanical events inherent in every activity within the ecosystem would become a macro-level problem in dealing with a system on this scale. Second, quantum indeterminacy must have been involved in the original creation of the ancestors of DNA 4 billion years ago, and has been a vital ingredient in the process of mutation ever since. Even if we ignored these factors and proceeded with our computer-based simulation, we will still confront tendencies to occur in the actual dynamics of organic life which display regularities that are not present in inorganic matter. The point is that we must in our study of biological life confront an event horizon of knowledge at which the harmonious interplay of the parts is not explained in terms of the collection of parts. And this is where profound complementarities inevitably emerge in modern physical theories.

One of the most promising recent contributions to the effort to better understand the distinction between organic and inorganic matter has been made by Ilya Prigogine, and is based on the study of thermodynamic processes.[27] What Prigogine and others have shown is that when a system is far from equilibrium, or where it is at a much higher temperature than its environment, new types of structure may originate "spontaneously." The result is that new dynamic states of matter, namely, organic life, are created. Prigogine calls these structures "dissipative"[28] to emphasize the role of dissipative processes, or of heat exchange, in their formation. Even though the second law of thermodynamics is not violated globally and entropy increases overall, these dissipative systems, or organic life forms, end up being more ordered at the expense of the environment which acquires a higher entropy. What this suggests is that there is a complementary relationship between organic and inorganic processes reflected in this fundamental distinction between dissipative and non-dissipative processes, or degrees of order and disorder measured in terms of entropy decrease or increase.

Prigogine has also made a major contribution in providing a basis for understanding the relationship between reversible and irreversible processes in a quantum mechanical universe. Reversible processes are viewed as independent of the arrow of time. In other words, you can run them backwards and arrive at the same points in the conditions of the system. Irreversible processes, on the other hand, depend on the arrow of time, and the order in which they run makes an enormous difference. It is the law of entropy increase in time which gives time its arrow, or a direction which distinguishes past from future. The laws of classical physics, although reversible, acquire an irreversibility when thermodynamics is brought into the picture. The laws of quantum physics, on the other hand, are not subject to this condition and remain reversible. The difference is easily illustrated—the aging of a piece of fruit is, in some sense, an illusion from the point of view of quantum physics. There is no suggestion of a progression from present, ripe fruit, to future, rotten fruit, on the quantum mechanical level. Irreversible time, or aging, makes no sense on this level. Yet the quantum mechanical system that is a piece of rotting fruit never appears to run backwards and become ripe once again, which would represent a violation of the law of increase of entropy. Since this distinction between quantum and macro-level processes appears to be rather categorical, it has often been appealed to in the effort to substantiate the claim that a categorical distinction between these two domains, and thus between classical physics and quantum physics, was valid in epistemological terms.

We can obviate the basis for this distinction altogether, suggests Prigogine, by understanding the relationship between reversible and time-reversal invariant processes in terms of the act of making measurements or observations in quantum physics. At the point at which we measure or observe a quantum system, it clearly becomes irreversible, and thus can be dealt with in terms of a progression in time from past to future. The possibilities given by the wave function reduce to those apparent in the act of measurement, and the system cannot run backwards or return to its initial state. Prigogine suggests that reversibility and irreversibility are complementary constructs, and that the constructs reveal themselves in the absence or presence of observation or measurement. Since the collapse of the wave function can be assumed to occur in all measurements or observations on the macro level, irreversibility and the arrow of time would obviously be apparent in the physics that seeks to coordinate experience on this level—classical physics. This could mean that classical physics does not have equal authority with quantum physics based on the assumption that they deal with two distinct domains of physical reality. It is rather that classical physics is an approximate description of the dynamics of physical reality in a quantum mechanical universe, and it is observation or interaction in this universe that explains the existence of irreversibility and the arrow of time in classical physics.

The Aspect experiments are nicely illustrative of the actual existence in nature of this profound complementarity. After measurements are made between space-like separated regions A and B, we discover correlations that appear simultaneous or instantaneous. If we assume the existence of some causal agent in time which carries a signal from A to B, which must be superluminar, we face a large quandary—relativity theory is not violated and light speed remains, therefore, the upper limit at which a signal can travel. The logical mistake, which complementarity serves to resolve, is that non-interaction with the system precludes interaction, and yet both constitute a complete view of the situation. Although we have interacted with the system with the detectors at the two points, the system that yields these results has not been interacted with "between" the points—it remains in a single, non-analyzable quantum state. That state, which as we have argued mirrors the wholeness of the cosmos, is not divisible into time categories, or any other categories, in the absence of interactions which allow the categories to become emergent in the form of phenomena.

What makes the element of time so peculiar in the Aspect experiments is that the construct itself is an emergent property of the system only when the system becomes irreversible in the act of measurement. In the absence of measurement, the system must be assumed to be in a reversible state. Since the reversible processes between the detectors have not been measured, they do not display the emergent property of time, and can only be characterized as "timeless." Where the processes become irreversible in the act of measurement, they enter time with the definitions of correlations between activity in time we call "simultaneous" or "instantaneous." If we try to force causal connections in time on the reversible processes that we have not measured and cannot measure, we are attempting to analyze the unanalyzable, and the exercise leads, predictably enough, absolutely nowhere.

This profound complementarity between reversible and irreversible processes could also provide a basis for comprehending a relationship that has seemed quite paradoxical to the symbol-making animal throughout his history—that between order and chaos. If we assume that order is identical with reversibility, and disorder or chaos with irreversibility, a generalization that makes sense in terms of mathematical physics, then order and chaos are complementary constructs in physical reality.[29] In this view we can no longer conceive of order and disorder as if both were seeking complete domination of the physical situation, or as if the two, like physical analogs of good and evil, were engaged in cosmic warfare. They appear rather as equally necessary and fundamental aspects of a seamless whole which displace one another as emergent properties in a particular situation. The recent discovery via computer simulations that there are chaotic solutions in the non-linear deterministic equations of classical physics is another case in point. The existence within the equations of classical physics of a chaos so complete that any hope of returning to unrestricted determinism seems utterly dashed makes perfect sense within the logical framework of complementarity. Any definition of order must also feature disorder as its complementary construct.

Although there are good reasons to believe that the logical framework of complementarity will, as Bohr put it, "allow us to achieve a harmonious com- prehension of ever wider aspects of our situation" by removing "apparent disharmony," will it open the glittering door behind which lies the meaning of meaning or the self-justifying rationale for our existence? Obviously not. The scientific description simply cannot, once again, capture or describe the essence that is the whole and has, therefore, nothing to say about its intent, purpose, overall design, etc. Even if we put epistemological and/or metaphysical questions aside, the scientific description functions something like a neutral screen upon which we can seemingly project our anthropomorphic wishes and desires endlessly without either validation or refutation by the description itself. In the next chapter we will try to further demonstrate the validity of the claim that the entire universe is a quantum system. We will also take this opportunity to expand upon the implications of the hypothesis advanced in this chapter—that complementarity is a fundamental structuring principle in constructions of reality in both ordinary and mathematical languages.

8
The Emergence of a New Vision: The Unfolding Universe

> In the world of quantum physics, no phenomenon is a phenomenon
> until it is a recorded phenomenon.
> —*John A. Wheeler*

The central thesis in this chapter is that we are moving toward acceptance of the proposition that the universe must be viewed as a quantum system not merely in its early stages but at all scales and times. If this thesis is correct, the epistemological situation confronted in the quantum domain, which the principle of complementarity describes and comprehends, must also be invoked in the study of the largest system known to us—the universe. The new epistemological situation that results could also serve, in our view, to resolve some now seemingly irresolvable observational problems associated with the big-bang models of the universe. Since this more narrowly scientific argument can best be appreciated by those with a background in astrophysics, we have chosen to place it in an appendix.

The primary ambition in this chapter will be to make the case that an acceptance of the proposition that we live in a quantum mechanical universe serves to explain why the logical framework of complementarity appears to be a fundamental dynamic in the constructions of human reality in both mathematical and ordinary language systems. What this view provides is a more reasonable and self-consistent basis for understanding why mathematical language, or more appropriately the language of mathematical physics, is more "privileged" than ordinary language in coordinating experience with physical reality. An acceptance of the proposition that we live in a quantum mechanical universe also requires that we recognize an unanticipated limit on the ability of mathematical physics to disclose or describe the behavior of the entire physical universe, or of physical reality-in-itself. And this could serve, for reasons we will explore, to provide a basis for a renewed dialogue between members of C.P. Snow's two cultures on the implications of the description of the origins and evolution of the cosmos provided by modern physical theory regarding the ultimate foundations of human consciousness and the relationship of this consciousness to the entire universe.

As we saw earlier, the resolution of the observation problem in quantum physics requires an awareness of the "prior" conditions existing in the mind of the observer. Since the same applies in our view to the resolution of the various observation problems in contemporary cosmology, we will begin with a brief history of modern cosmology with the intent of illustrating the manner in which classical assumptions have functioned as "distorting lenses" in this area of physics as well. The other parallel with quantum physics apparent here is that it is the "bounded" progress of scientific knowledge which has allowed us to disclose the distorting lenses in this area of science as well, no matter what violence has been done to our linguistically-based and culturally derived conceptions of the way things are.

A Brief History of Modern Cosmology

Newtonian dynamics coupled with the law of universal gravitation allowed the motions of the planets and the stars to be understood for the first time within the context of physical theory. Although classical physics suggested that the universe was much larger than had been previously assumed, it could not say anything about its origins. The assumptions which Newton made about the large-scale structure of the universe, which were largely motivated by aesthetic and metaphysical concerns, were that it was infinitely extended and static, or eternally consistent. The aesthetic and/or metaphysical appeal of the idealization of the universe as static or eternally the same also seems to have carried over into Einstein's first general relativistic model of the universe. The universe in this model was conceived as finite, spherical, and static. When Einstein published his general theory of relativity in 1915, he did not anticipate that it would soon become the basis for positing the existence of a non-static, evolving model for the universe.

The first convincing evidence that the universe was expanding and evolving came from observational astronomy. New instruments probing deep space, such as the 100-inch Mt. Wilson telescope in the 1920s and the 200-inch telescope at Mt. Palomar in the 1940s, broadened the horizon of the observable universe to thousands of millions of light-years. Using these instruments observational astronomers, most notably Edwin Hubble, gathered evidence in the 1920s which strongly indicated that the universe was expanding and evolving. With the theoretical framework of general relativity already in place, the Belgian cosmologist Abbé Lemaitre and the Russian mathematician Alexander Friedmann postulated early in the 1920s what would have seemed preposterous a few years before. They advanced a model of the universe which was dynamic, expanding and evolving. These early relativistic models were, however, premised on a set of assumptions, reflected in the so-called Cosmological Principle, which were essentially an extension of Newtonian concepts of the universe. This principle states that the universe must be isotropic, meaning the same in all directions, and homogenous, meaning that the density of the universe is the same, on average, at all points in space. In contrast to Newtonian concepts, however, as the universe expands, the average density of matter in the universe over time would get lower, and the universe would become more dilute.

The appeal of the construct of an eternal universe is also apparent in the steady state theory advanced by Herman Bondi, Thomas Gold, and Fred Hoyle in the 1940s. In this theoretical model we have an expanding universe which, nevertheless, appears the same to all observers at all times because it obeys the Perfect Cosmological Principle. The assumption here is not only that the universe is homogenous and isotropic, but also that it looks the same to observers at "all" times. Although the model accommodated the observational evidence supporting an

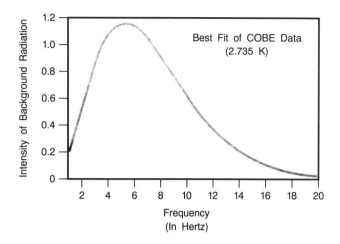

FIGURE 20. Data from COBE showing that the cosmic background radiation fits remarkably well a black body curve at 2.735 K (*see also Science News*, 1990, *137*, p. 36).

expanding universe, it did so within the framework of an eternal universe. One large price that had to be paid to accomplish this was the abandonment of a cherished foundational principle in physics—the conservation of mass and energy. Matter in this model is assumed to be created continuously ex nihilo. As the universe expands, space is filled with newly created matter, and thus the average density of the universe always remains the same. For this reason the steady state model is also termed the continuous creation model. What these ad hoc assumptions allowed one to do was to avoid, or obviate, a central question—how did the universe come into existence? If an evolving universe without continuous creation requires that the average density of the universe must get lower over time, then running the clock backwards would imply that the universe must have been much denser and hotter much earlier in time. As Lemaitre, Friedmann, and others reasoned in the 1920s, this would also suggest that the process could eventually be traced back to a beginning point, or a primordial singularity.

In the early 1950s, the cosmologist George Gamow extended Lemaitre's and Friedmann's original ideas. At this point quantum physics was well established, and could begin to provide some insights into the early life of the universe where energies and temperatures associated with the domain of violent nuclear collisions would be apparent. If the label for this model which stuck, the "big bang," seems less than dignified in describing how the grandeur of the cosmos came into being, there is a reason why this might be the case. The term "big bang" comes from Fred Hoyle, who advocated the opposing steady state model, and was not intended to dignify Gamow's theory. Yet it was not clear for several years which of the two views would triumph. Although the steady state theory implied a theoretically simpler universe, observational astronomy, armed with the new branch of millimeter and radio astronomy and sophisticated optical spectroscopy, provided

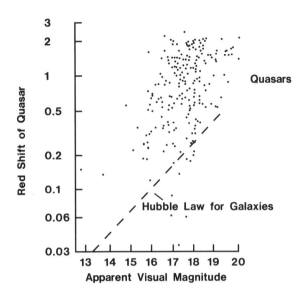

FIGURE 21. The Hubble Diagram of quasars showing how the speed of recession of a quasar measured by its redshift correlates with the distance of the quasar. Quasars do not obey the law followed by galaxies. Data collected by astronomer Dr. E.M. Burbridge (*see* M. Berry, *Principles of Cosmology and Gravitation*, Cambridge: Cambridge University Press, 1976).

evidence that was strongly in favor of the big-bang theory. The two pieces of experimental evidence that produced this result were the discovery of the 3-degree Kelvin black body radiation, and a new class of astronomical objects known as quasars.

Recently, NASA launched the *Cosmic Background Explorer*, a satellite designed to accurately measure the temperature of the cosmic background radiation and any departures of this radiation from perfect isotropy. The results, which are very surprising, indicate that the universal radiation is extremely smooth and follows to an extremely high degree of accuracy a black body fit at T = 2.735 degrees Kelvin.

The existence of the 3-degree K (more accurately 2.735 K) microwave black body radiation, which was accidentally discovered by Bell Telephone physicists Arno Penzias and Robert Wilson in 1965, had been predicted by Gamow for some time. Gamow's assumption was that a background of radiation that fills all space would represent the relic radiation from the initial big bang and would fill a greater volume of space as the universe expands and cools. Since the total energy content of the universe is conserved, the energy density, or amount of energy of radiation per unit volume, must go down. This means that what used to be a hot background of radiation, composed of X-rays and gamma rays, would be redshifted to the microwave region of the spectrum. In other words, as the universe expands the background radiation would acquire a much lower temperature. Since the big-bang theory could explain the existence of the primordial radiation, and even served to predict its existence, it appeared more victorious in competition with the steady state theory. Steady state does not allow a beginning, and the microwave background within the context of that theory must presumably arise from many distant sources that add up together to produce an unresolved background of radiation. Yet no one has been able to discover suitable candidates for these alleged sources.

The second major blow to steady state theory was the discovery of the quasars in the 1960s. A remarkable feature of the quasars is that these objects are extremely distant from earth. In order to appreciate how distant these objects are, we must first remind ourselves that the maximum distance from earth where receding galaxies can be observed is a function of the speed of light. If the speed of recession of celestial objects becomes equal to the speed of light, these objects simply cannot be observed from earth. This limit on observation is known as the horizon of the universe. The distance to other galaxies that can be seen within this horizon is determined by the extent to which light from a galaxy has been redshifted, or has moved to the red part of the light spectrum. Thus the light in spectral lines of elements emitted by a more distant galaxy will be more redshifted than that from a nearer galaxy. Although often explained by appealing to the well-known Doppler effect which occurs when, for example, one detects the shift of the sound of a receding siren to lower pitch, the redshift of light in an expanding universe is a general relativistic result arising from the stretching out of space-time. It was first observed by Hubble, and is the basis for what we now call the Hubble law—the more distant a galaxy, the more its spectrum is shifted to lower frequencies or longer wavelengths.

What was unusual about the light of the spectral lines from the quasars or QSOs, meaning "quasi-stellar objects," is that this light has been redshifted to very low frequencies, as compared with light from other celestial objects. The redshift for quasars is so high compared to other astronomical objects that we must presume that these objects are near the edges of the observable universe, or at the 80% – 90% level of the distance to the horizon. Some quasars are so distant from us that they seem to be receding, or moving away from us, at speeds exceeding 90% of the speed of light.

Eventually it was understood that quasars are distant galaxies, and that the light from these galaxies has been traveling for billions of light-years. Since light traveling such distances normally becomes so faint that the source cannot be observed, it was also understood that the brilliant, star-like nucleus of the ancient galaxies from which the light of the quasars originated must have been far more brilliant than the nucleus of nearby galaxies. What this suggests is that the universe does not, in fact, look the same at all times to all observers. If galaxies were much brighter in the past than they are today, then the Perfect Cosmological Principle, which forms the foundation of the steady state theory, does not hold. The apparent disagreement of steady state theory with observations finally prompted even its original proponents to abandon it soon after the discovery of QSOs. But what they did not anticipate was that the big-bang theory would itself face some challenges as improved cosmological observations became available.

The most significant of recent developments in theoretical cosmology is that advances in quantum field theories have afforded us a unique opportunity to seriously attempt to describe the conditions of the universe close to the big bang. This was not possible until recently because the extremely high temperatures and densities prevalent in the early universe required answers from particle physics which were totally unknown. In our new situation particle physicists and cosmologists have combined their efforts, and can speculate upon conditions in the early universe, or at least conditions after the first three minutes, that can be increasingly tested in high speed particle accelerators. There is now general agreement that the very early universe, barely a fraction of a second old, was in a quantum state, and that a quantum theory of gravity is the central ingredient in moving toward a more complete and self-consistent cosmology. Progress in this area over the last few years has been quite stunning due largely to the success of the quantum field theory of particle interactions.

Problems with Big-Bang Theory

The three large problems faced by all big-bang models are known as the flatness problem, the horizon problem, and the isotropy—and homogeneity—problem (see Plate III, p. xii). The flatness problem has to do with the fact that all of the original

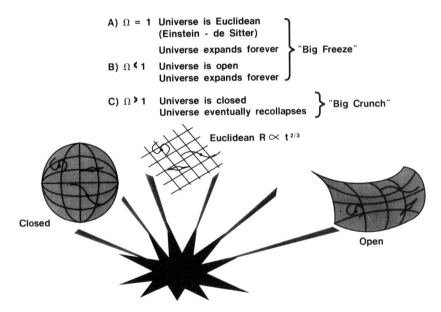

FIGURE 22. The geometries of the three types of big bang cosmological models, open ($\Omega < 1$), closed ($\Omega > 1$), and flat ($\Omega = 1$).

big-bang theoretical models cannot account for the observed density of matter in the universe. What is curious about this observed value is that it is quite close to what would be required to bind the universe. Quantitatively this is expressed as Ω ~ 1, where Ω is the density ratio $\Omega = \rho/\rho_{crit}$ and the critical density can be expressed as[1]

$$\rho_{crit} = 2 \times 10^{-29} \, (H_0/100 \text{ km/sec/Mpc})^2$$

where H_0 is the so-called Hubble constant which defines the rate of expansion of the universe. If $H_0 = 100$ km/sec/Mpc, it means that as the distance increases 1 megaparsec (Mpc), or 3 million light-years, the rate of expansion of the universe increases by 100 kilometers per second. For $H_0 = 100$ km/sec/Mpc, the horizon of the universe lies 10 billion light-years away from the earth.

Although the precise value of H_0 remains the most fundamental challenge of observational cosmology, most observers presently believe that it lies in the approximate range 50 – 100 km/sec/Mpc. In terms of Friedmann's general relativistic models, this implies that the horizon is 20 – 10 billion light-years away, and that the universe is approximately 20 – 10 billion years old.

If Ω turned out to be precisely equal to unity, the geometry of the universe would be exactly flat, and the universe would expand until it finally comes to rest in the infinitely distant future. If the density turns out to be less than the critical value, the geometry of the universe would be similar to the surface of a saddle. This would mean that the universe is open and infinite in extent, and, like the flat universe, that its expansion would never cease. Suppose, however, that the density turns out to be greater than the critical density. The geometry in that event would resemble the surface of a sphere, and the universe would, therefore, be closed. In this case, the universe would expand to a maximum size, and then start recollapsing back onto itself in the so-called *big crunch.*

The mean density of the universe is computed by estimating the masses of distant galaxies, based on their observed brightness, and comparing those values with what is known for nearby galaxies. When all the luminous matter in a given volume of space is added up, the density of luminous matter is determined. The obvious problem is that current observations cannot unequivocally distinguish the type of universe we live in. Although most observers favor values close to 0.1, values of Ω for luminous matter are in the range 0.1 to 2. If the only type of matter that there is in the universe is luminous matter, like that found in stars and nebulae, the present results indicate the universe is open. Astronomers suspect, however, that there is much more dark matter in the universe than all the luminous matter making up the galaxies. It is conceivable that more than 90% of the matter in the universe is in the form of dark, invisible matter. Yet attempts thus far to discover this "missing mass" have not been successful.

Although present observations indicate only an approximate range of the mean density of the universe, what is perplexing is that the value is so close to the critical density required for a flat geometry. Even if it turns out that the universe is not quite flat today, the universe in its early stages must have been incredibly flat[2], to

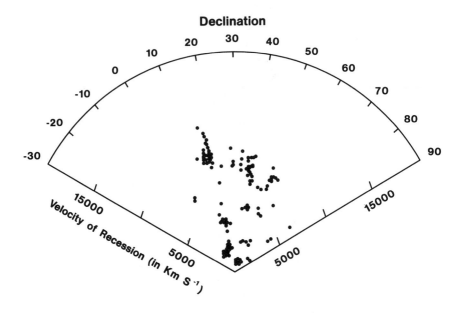

FIGURE 23. The Hercules Supercluster is a large structure of many galaxies which seemingly violates the condition of homogeneity of matter in the universe. The other celestial coordinate, right ascension, is restricted in the range 15 h 30 m and 17 h (Data can be found in J.O. Burns, J.W. Moody, J.P. Brodie, and D.J. Batushi, *Astrophysical Journal*, 1988, *355*, p. 542).

one part in 10^{50}. The reason for this conclusion is straightforward. In the absense of inflation, if the universe, which has expanded by 60 orders of magnitude since the big bang, is approximately flat at present, this would translate into a condition of rather exact flatness in its early stages.

The horizon problem relates to the uniformity of the 3-degree K black body radiation—it has the same temperature to within one part in ten thousand in every direction on the sky. The problem is that the original big-bang model indicates that opposite parts in the sky at the time that the microwave background formed some 10^5 years from the beginning were separated by distances of 10^7 light-years. Given the observed uniformity of temperatures in this radiation from all parts of the sky, one would have to conclude that opposite parts of the sky were in causal contact. It is noteworthy here that all big-bang models, including the inflationary scenario, presume direct causality. Yet the theory of relativity insists that no signal can travel faster than light, and that opposite parts of the sky were, therefore, space-like in their separation.

The last problems that the original big-bang model cannot explain is the isotropy and homogeneity problems. Given the large number of possibilities in the initial conditions allowed by the model, one would not expect the universe to be expanding in a rather smooth fashion today. Moreover, assuming that the universe started from a hot big bang, one would expect that the total chaos that prevailed in the early instances would not have died away and that the universe would be expanding in a very irregular fashion today. Indeed, quantitative calculations show that slight variations, or inhomogeneities, from the mean uniformity of matter would have been greatly amplified. Since these variations should be reflected in some variation in the observed isotropy of the background radiation, NASA launched COBE, designed to detect the existence of the background radiation. Although the view of the universe as isotropic and homogeneous was considerably reinforced by the discovery that the background microwave radiation appeared remarkably uniform no matter where one examines it in space, we have also discovered that matter appears to be highly non-uniform in its past distribution. The original expectation was that at some point in deep space we would begin to observe uniformity, or homogeneity, in the density of matter analogous to that which had been observed in the background microwave radiation. The problem is that this does not appear to be the case even out to distances of more than 400 million light years. There is, for example, no such uniformity or homogeneity of matter in the superclusters which are hundreds of millions of light-years distant from us.

What seems clear at this point is that the universe is not homogeneous in its matter distribution at all scales and in all directions. On the other hand, radiation, as the recent COBE results indicate, is extremely isotropic. As astronomers look further and further into space, some incredibly large structures have been observed. One such structure, known as the Pisces-Cetus supercluster, extends to a few percent of the observable radius of the universe.[3] Galaxies now appear to cluster themselves in increasing hierarchies of clusters, called superclusters, and may even form larger structures, or super-superclusters, which approach a sizable fraction of the radius of the observable universe. If gravity is the only force operating at large distances and determining the overall structure of the universe, time scales much longer than the age of the universe would be required for these structures to grow to their present sizes. Their very existence seemingly contradicts the big-bang theory itself. Often these large superclusters assume the shape of filaments which appear to lie on the surface of very large bubbles of empty space termed the "voids." Also, galaxies have been observed which do not move in the isotropic fashion required by the uniform expansion of the universe.[4] What is perhaps more surprising, based on the totality of what is known in contemporary cosmology, is that the universe is not more anisotropic in matter than we have observed it to be.

Inflation as the Solution to the Big-Bang Problems

The current big-bang model attempts to eliminate the flatness problem, the horizon problem, and the isotropy problem with the addition of the so-called *inflationary hypothesis.* Whether the inflationary model can predict the observed structure of matter in the universe and explain why the universe remains inhomogeneous at all scales that have been observed so far remains an open question. Moreover, the inflationary model exacerbates the resolution of a well-known theoretical problem of current physics—why is the cosmological constant so small? Current theories of elementary particles and fields indicate that the vacuum should have a nonzero energy density. Nevertheless, observations of the large scale structure of the universe indicate that such a nonzero energy density would have to be smaller than current particle physics theories indicate by a staggering factor of 10^{120} or 1 followed by 120 zeroes!

Leaving these problems aside for the moment, the inflationary hypothesis presumes that the universe comes into existence out of the vacuum state, or out of a completely unvisualizable "point nothing" capable of generating enormous energy. In the first phase, or at the initial singularity, the space-time description breaks down entirely. The expectation here is that a proper quantum treatment of gravity could eliminate the theoretical problems associated with a singularity. This initial quantum gravity phase is allegedly followed by the "inflationary era" at about 10^{-35} second.[5] During the alleged inflationary era, the universe presumably underwent an extremely rapid expansion, doubling in size every 10^{-35} of a second, eventually expanding in size by a staggering factor of 10^{50} or more.

The process through which water freezes provides a crude analogy for understanding inflation. Like the freezing of water, inflation involves a phase transition. Prior to inflation, the universe is viewed as being in a phase of symmetry with respect to the so-called Higgs fields, and the strong, weak, and electromagnetic interactions were unified. The Higgs fields are members of a special set of quantum fields postulated in Grand Unified Theories (GUT), and they serve to account for the spontaneous symmetry breaking that would lead to their emergence. The state of unbroken symmetry when the Higgs fields were zero, or prior to the onset of inflation, is known as the false vacuum, which is alleged to have contained within itself tremendous energies. The numbers here are quite impressive—every cubic centimeter of volume in this false vacuum would have to contain 10^{95} ergs of energy, or about 10^{19} more energy than the mass converted into energy of the entire observable universe.

As this early universe expands and cools to a temperature of about 10^{27} K, the inflationary hypothesis assumes that a phase transition begins from the false vacuum, where all the Higgs fields were zero, to a less energetic phase termed the "true" vacuum. In this true vacuum state, the Higgs fields acquire non-zero values, and the GUT symmetry breaks down. At this point the universe enters an energetically stable phase, meaning that the amount of energy available to it for the remainder of its existence was fixed. The breaking of the GUT symmetry in the expanding universe allowed the strong force to separate from the electroweak force.

One of the peculiar properties of the false vacuum during the GUT era is its pressure. In normal gases, the pressure is positive, and an aggregate of hot gases, like that of a star, would tend to expand while gravity would tend to pull it together. The false vacuum during the inflationary era is understood, however, as having a very large negative pressure. Although there was no ordinary hot matter and no ordinary pressure, we are obliged here to conceive of a tremendous negative pressure coming out of the nothingness of the false vacuum. As a result, the equations of general relativity predict that the early universe, rather than being pulled together by gravity, would be pushed away. Thus the universe inflates.

The inflationary model requires that we view the universe as expanding over a period of 10^{-30} second from a size of less than 10^{-50} of a centimeter to about one centimeter. This represents more than fifty orders of magnitude! The subsequent expansion of the universe to its presently observed size of some 15 billion light-years or more is trivial when compared to the staggering expansion called for in the inflationary model. As the universe was moving from the nothingness of the false vacuum to the nothingness of the true vacuum, the tremendous energies locked up in the negative pressure of the false vacuum were allegedly released. The inflationary model not only requires that we view all of the observable matter in the universe as coming out of the nothingness of the false vacuum; it also implies that the actual size of the universe should be much larger than what we can observe from the earth. The big-bang model with inflation requires a horizon that is considerably larger than the size of the universe The universe required in this model would, in fact, have to be 10 million billion times larger than we observe it to be.

What is most important to realize about the inflationary model is that it was not originally proposed for any compelling theoretical reasons. It arose principally in the effort to solve the three observational problems faced by the standard big-bang theory. Inflation seeks to solve the flatness problem by assuming that flatness would be a natural consequence of rapidly inflating the curvature of space-time in the same way that wrinkles on the surface of a spherical balloon get smoothed out as the balloon gets inflated to larger sizes. No matter what curvature one starts out with, the expansion of space-time by a factor of 10^{50} or more would create a universe that was remarkably flat. If one, for example, were to inflate a balloon by such a factor, the surface of this balloon would closely approximate a flat surface. The horizon problem is solved by this model in that it allows us to assume that regions on the opposite side of the expanding universe were in causal contact after the singularity and before they were blasted apart by inflation. Inflation has not yet solved the extreme isotropy problem for radiation, and there are some recent indications that it will face serious problems in the effort to do so. Moreover, it fails to account even more for the large-scale non-homogeneity of matter distribution in the face of the extreme smoothness of radiation.

The remainder of the big-bang story after the inflationary phase story is less controversial. Between 10^{-30} and 10^{-6} second, the universe was filled with a primordial soup of quarks and light mass particles, like the electrons. After about 10^{-6} second from the big bang, the quarks combined together to form the heavy particles, like the protons and the neutrons. In the next phase of ascending complexity, between 1 second and 3 minutes, protons and neutrons undergo nuclear reactions forming nuclei of helium and its isotopes. Although some heavy elements were formed during this period, their amounts were minute. The formation of appreciably large amounts of those elements occurred millions of years later in massive stars that evolved into exploding supernovas. The heavy elements were predominantly created in the nuclear reactions of supernovae and spewed outward into space by this most powerful type of all stellar explosions.

Until about 100,000 years after the big bang, photons and matter were coupled together, and the universe was dark, or opaque to its own radiation. The 3-degree K black body photons escaped after this time. What is important to realize at this point in the story is that these photons provide no opportunity to probe periods prior to 100,000 years where fundamentally important quantum physical processes were taking place. This means that the main tool of observational astronomy, light quanta, simply cannot be used to verify what occurred during this crucial early quantum phase of the universe. The other means that we have of probing this period, like estimating the amount of primordial helium, are not as direct. What we have here is a situation in which the limits that observations impose upon the theory are intrinsic—they will not, in other words, go away with future improvements in observations. Moreover, it is not even clear that the theoretical foundation of inflation, GUT itself, will prove to be a viable physical theory.

The Universe as a Quantum System

Our argument is that the observation problems associated with the big-bang model might be eliminated by viewing the universe as a quantum system (see Appendix). Although this assumption is more in accord with the totality of what is known about the universe, the proposed solution is not as simple as it might first appear. If the universe is a quantum system, it is obvious, first of all, that we can no longer treat it as a closed system in the sense of being separate and discrete from the observer. Observations of even the largest system known to us would also include the observer and his measuring instruments. Although this view suggests that all acts of observation in cosmology should be theoretically subject to this condition, we will confine ourselves here to discussing those in which the rules of observation in quantum physics must clearly be invoked. Since what we are proposing represents a radical revision of normative assumptions in cosmology, we will first seek to explain why cosmologists in our view have resisted the assumption that the universe is a quantum system.

The primary sources of this resistance are, we think, similar to those discussed earlier in connection with alternatives to the Copenhagen Interpretation. Although quantum physics has been quite successful in describing conditions in the early universe, cosmologists and particle physicists have, in general, resisted the notion that what was clearly a quantum system in the early stage continued to manifest as such in the later stages. Although quantum theory is widely accepted as the only means of arriving at a description of the very early stages in the life of the universe, the description of all subsequent stages in the big-bang model with inflation is premised on assumptions that come from classical physics. The tendency has been to assume that the early universe must have been a quantum system, and then to studiously avoid the implications in terms of the role of observer and the continuity of physical processes in the treatment of all subsequent stages.

The largest single explanation as to why physicists have been able to continue to comfortably operate in this way is that the general theory of relativity has not yet been quantized. Since the general theory of relativity describes the large-scale structure of the universe in terms of the force of gravity, it applies to scales a few miles across to scales the size of the observable universe. Quantum mechanics, on the other hand, is normally presumed to apply only in our dealings with nature on extremely small scales, such as a millionth of a millionth of an inch. Since cosmologists appeal to both theories in efforts to describe the origins and evolution of the cosmos, there is the understandable tendency to presume that they need not introduce quantum physics, or confront the epistemological problems associated with quantum physics, on the large scales where the general theory applies.

Although the general theory has, of course, been highly successful, it is still basically anchored in classical terms. The conceptual framework of general relativity, the space-time continuum, is, as the name implies, a continuum. In Einstein's original theory, there is no place for discrete quantum jumps and interactions. Although a number of recent attempts have been made to quantize gravity, the mathematical descriptions when the proposed quanta of gravity, the "gravitons," are introduced contain, in contrast with quantum electrodynamics, too many infinities to be workable. The manner in which a general relativistic description of the expanding universe arises out of a quantum mechanical description, which is fundamentally different, remains, therefore, a great enigma. Yet if a quantum theory of gravity removes this enigma, which is what is anticipated here, cosmologists will not only be obliged to contend with the prospect that the entire universe is a quantum system; they should also be forced to confront problems of quantum epistemology in the study of the universe.

Another reason, in our view, why physicists have not been inclined to view the entire universe as a quantum system is more directly related to the legacy of classical physics. In classical physics, as we have seen, parts can theoretically be known with absolute certainty, and the ultimate assemblage of those parts constitutes the whole. Since the classical physicist believed in locality and unrestricted causality, the part could be treated as a closed system separate and distinct from the observer. Relying on the laws of conservation of energy, momentum, and angular momentum, classical concepts allowed a cosmologist to presume that he could know with ultimate certainty the actual state of a collection of parts in a larger and similarly closed system. And the system representing the ultimate assemblage of parts constituting a whole was the universe itself. The assumption of unrestricted causality also led to the presumption that subsequent events in larger and larger systems could be known to an arbitrarily high degree of accuracy after an observation disclosed the initial conditions in the systems. It was, therefore, reasonable to assume that if the description of the initial conditions in the largest system that can be known, the universe, was sufficiently complete, then we could presume knowledge of all subsequent events in the evolution of the universe. Since knowledge of the universe in this sense is dependent upon the existence of a one-to-one correspondence between every aspect of the physical theory and the physical reality, cosmologists are understandably reluctant to accept the view of the cosmos as a quantum system and to confront the problem of quantum epistemology.

Although it seems quite clear that many cosmologists, wittingly or unwittingly, operate on a belief in this correspondence in the search for an experimentally verifiable theory for the origins of the cosmos, they are applying classical assumptions in a situation where they do not by any reasonable criteria make sense. Quantum effects in the very early stages of the life of the cosmos were large, and the most modern of all cosmological theories—the big-bang model with inflation—speculates that the universe first came into existence as a result of a quantum transition.[6] During the so-called Planck era close to origins or the singularity, which begins at times earlier that 10^{-44} second, quantum effects would be pervasive and inescapable. In the Planck era, even the fundamental construct of space-time is not applicable, and what we are asked to envision here is a universe filled with mini black holes quickly arising from and disappearing into a background of nothingness. The conditions of that era are still present today for sizes approaching the Planck length, or in the unimaginably small dimensions of 10^{-33} cm. Theoretically at least, we could observe effects present at the singularity if we could generate sufficient energy to simulate earlier conditions.

But even if we could conduct this experiment, which would require that we generate energies well beyond any technical capabilities now imaginable, would the information provided about initial conditions allow us to infer approximate understanding of all subsequent events? Obviously not. This is a quantum domain requiring quantum experiments and rules of observation. Here the quantum of action definitely makes the situation far too ambiguous to presume accurate prediction of all future events. In other words, we have an observational problem which exists in principle irrespective of future improvements in observations. If we insist on applying classical presuppositions, our experience in quantum physics clearly suggests that this will only lead to irresolvable ambiguities. Yet dealing with the observation problem in a quantum universe is, as John Bell has noted, more than a little problematic. In a laboratory setting, the requirement that we factor into our understanding of results in quantum physics the relationship between the observer and the observed system can be met. But how does one meet that requirement when the observed system is the entire universe? This problem becomes particularly acute in a practical or operational sense in the very early stages of the life of the universe where quantum effects are extremely important.

Yet our situation indicates that this problem cannot be ignored. Quantum effects, however small they might be on the macro level and however ignorable they might be as a matter of convenience, are pervasive throughout the history of the universe. It is, therefore, quantum theory that promises to provide the complete description of the history of the physical evolution of the cosmos. One cannot, in theory at least, ever presume a categorical distinction between acts of observation and the observed system, even if the system is in excess of the billion galaxies contained in the observable universe. We, as observers, are clearly included in experimental situations dealing with the largest system imaginable just as we are included in dealing with the smallest possible system in the quantum domain. The latter is implicit in the former, and any theory which does not take into account quantum effects is dealing, in our view, in higher order approximations. The only way in which we could conceivably know the history of the universe in the classical sense would be to know the sum of all energies, momenta and other physical quantities of all objects in the universe at any one instant including those present at origin. Quantum indeterminacy obviates that prospect in principle. There is, therefore, no outside perspective from which to view the physical universe, and any theory based on such a perspective is, in our view, an idealization that cannot be in the final analysis self-consistent.

The Observational Problem in Cosmology

Although most cosmologists view the early universe as a quantum system, they continue, as we have suggested, to treat its evolution in classical terms. This also means that they are obliged to view the resolution of observation problems in terms of classical assumptions about the independent existence of macroscopic properties of the observed system—in this case the entire universe. The usual assumption here is that quantum effects in observations pertaining to the large-scale structure of the universe as provided by the existence of superclusters are not large enough to have any substantive impact on the accuracy of these observations and can, therefore, be safely ignored. Yet this argument is seriously flawed in our view for the following reasons: As we look deeper and deeper into space in our efforts to understand the large-scale structure of the universe, photons, or light quanta, which are our principal source of knowledge about phenomena in the early universe, become fewer and fewer in number. Theory presumes that galaxies exist beyond the current observational limits of redshift $z \sim 3\text{--}4$. However, in order to observe these galaxies, one needs accurate determination of redshift or the associated distance. The specific observational choices prohibit this from happening in practice because observational limits inherent in dealing with small numbers of quanta emitted from galaxies at large z do not allow an accurate determination of *both* position *and* distance of galaxy in a *single* observation. This means that we are eventually observing photons on our telescopes which are so few in number that wave-particle duality, and thus quantum uncertainty, could have substantive consequences in making acts of observation. In these situations the "choice" of whether to record the particle or wave aspect could have appreciable consequences. This, we believe, may turn out to be the cause. The photographic evidence produced by these observations involves the particle aspect of light quanta, and observations based on spectral analysis involve the wave aspect of these quanta. If we are observing only a few photons from very distant sources at the edge of the observable universe, it would seem that the resulting indeterminacy would be imposing some limits on the process of observation.[7] [See Appendix] Our view is that *complementarity will have to be invoked in our efforts to understand the early life of the universe based on observations involving few light quanta due to the irremedial ambiguities introduced by the quantum of action.*

If the observational choice has unavoidable consequences which result in logically different views of the universe, perhaps what is being disclosed here is an *inherent* horizon of knowledge in our attempts to describe the universe. What makes this hypothesis reasonable is the fact that observation at the point at which this inherent horizon appears clearly invokes the rules of observation in quantum physics. This horizon appears as we attempt to detect fainter and fainter sources in the sky, and thus we will ultimately be obliged to deal with fewer and fewer photons. Since observation involving small numbers of photons clearly invokes the quantum measurement problem, the choice of which aspect of the quantum reality we elect to observe might have appreciable consequences.

The only way in which we can presume to know all the macroscopic variables involved when the observation is based on small numbers of photons is to treat what is clearly a quantum system as a classical system. When fewer and fewer photons are detected from increasingly fainter sources, it seems unavoidable that the wave-particle complementary nature of quanta should be invoked, and that this aspect of quantum reality must be factored into our understanding of the experimental results. The apparent fact that observation in such situations suggests that we must invoke wave-particle duality becomes particularly intriguing when we consider that observational limits appear to be a consequence of adopting "single" and "specific" theoretical models of the universe. Why is it, for example, that the big-bang model leads to ambiguities which cannot apparently be resolved within the context of that model? Or why is it that any single model of the universe seems to inevitably result in ambiguities that seem impossible to resolve through observation? The usual answer to these questions is that improved theory coupled with improved observation will resolve the ambiguities. Yet there are now good reasons to believe that the observational limits, or what we term horizons of knowledge, are inherent in the laboratory we call the universe, and that improved observational techniques simply cannot remove the ambiguities.

The radical suggestion here, which will doubtless disturb many cosmologists, is that these horizons of knowledge are testifying to the fact that the observed system is a quantum system. If this is the case, then it is conceivable that single and specific models create ambiguities which cannot be resolved within the context of those models because we have failed to consider the prospect that logically disparate models could be complementary. If the horizons of knowledge present in any cosmological model prevent us from unequivocally confirming or rejecting a particular theoretical model[8], this could be because the observed system is a quantum system. Our suggestion is, therefore, that our problem is not as much empirical as it is conceptual. If we assume that the universe is a quantum system, perhaps it is an inevitable outcome of the process of observing fewer and fewer quanta from the early universe that we should confront profound complementarities. If the observational limit is, as it now seems to be, due to inherent uncertainties endemic to all acts of observation in the quantum domain, that limit could be introducing us to additional complementarities in the form of logical disparate cosmological models which, taken together, define the entire situation. [See Appendix]

FIGURE 24. Wheeler's gravitational lens experiment utilizes a gravitational galactic lens to perform a "delayed-choice" experiment. Depending on where one places the light detector, one determines "now" what path or paths (option A: both paths; option B: one path) the photon "took" on its way to the earth.

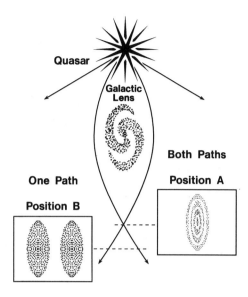

Non-locality and Cosmology

If we are correct in assuming that the entire universe is a quantum system and must be regarded as such in cosmology, this new situation could entice us into drawing some dramatic conclusions in epistemological terms. What the new epistemology requires, as we have argued throughout, is overcoming the bias that we are separate from the observed system and the associated doctrine of a one-to-one correspondence between every element of the physical theory and the physical reality. We have made the case that Bell's theorem and the experiments testing that theorem have forced us to embrace a profound new synthesis between the observer and the observed system by confirming the real or actual existence of non-locality as a new fact of nature. Another experiment which carries large implications in the effort to understand our proposed new view of the epistemological situation in cosmology is the delayed-choice experiment of Wheeler we discussed earlier.

Recall that Wheeler's original delayed-choice experiment was a thought experiment which has since been translated into actual experiments. What these experiments illustrate is that our observations of past events seem to be influenced by choices that we make in the present. In our view, what the experiments are validating is the actual existence of another type of non-locality. In the delayed-choice experiments, not only does the collapse of the wave function occur over any distance—it also seems obvious that it occurs in a way that is insensitive to the arrow of time. The second type of non-locality which we must also now view as a fact of nature is, therefore, temporal non-locality. More important, both spatial and temporal non-localities provide the basis for conceiving of the space-time complementarity in a way commensurate with the quantum paradigm.

In order to better illustrate the importance of temporal non-locality in making cosmological observations, let us consider the following ingenious delayed-choice experiment devised by Wheeler. In this experiment light emitted from a quasar passes by an intervening galaxy which serves, in effect, as a gravitational lens. The light can be presumed to be traveling in two paths—in a straight line path from the quasar and a bent path caused by the gravitational lens. Inserting a half-silvered mirror at the end of the two paths with a photon detector behind each mirror, we should arrive at the conclusion that the light has indeed followed the two paths. If, however, we observe the light in the absence of the half-silvered mirror, our conclusion would be that the light traveled in only one path and only one detector would register a photon. What we have done here is chosen to measure one or the other complementary aspect of this light with quite different results. The insertion of the half-silvered mirror, as in the laboratory version of the delayed-choice experiments, allows the wave aspect of this light to manifest as reflection, refraction, and interference. The removal of the mirror registers, in contrast, only the particle aspect.

What is dramatic in this experiment is that we would be apparently determining the path of light traveling for billions of years by an act of measurement in the last nanosecond. In accordance with the Copenhagen Interpretation, however, conferring reality on the photon path without taking into account the experimental setup is not allowed. This reality cannot be verified as such in the absence of observation or experiment. What is determined by the act of observation, as we understand it, is a "view" of the universe in our conscious construction of the reality of the universe, and the "existence" of the reality-in-itself is not legislated by this view. In other words, although our views of reality are clearly conditioned by acts of observation, the existence of the reality itself is not in question. What seems quite clear in this experiment is that we cannot take the reality of the photon path for granted[9], and our observational "view" in the present does have an irremedial influence upon our "view" of the past. It also seems clear that this view would be necessarily ambiguous if we do not take into account the manner in which it is conditioned by the act of observation.

Yet what astronomers normally do here is confer reality on the photon path implicitly without paying any heed to the means of observation. We should emphasize that in this particular situation factoring out the observational limits that might have been occasioned by quantum indeterminacy does not eliminate this problem. What we have here is an epistemological problem which cannot be circumvented by any appeal to "practical" solutions. Also, if we presume that the collapse of the wave function occurs with any quantum transition, which seems more reasonable and certainly less anthropomorphic than viewing the collapse as a consequence of measurement by intelligent observers, then our view is conditioned by observational limits that are incalculably large in any given observation.

Viewing this problem in its proper context, however, we are not driven to the conclusion that we should give up making astronomical observations. What we are obliged to do rather is re-examine the experimental situation in terms of what it clearly implies about the new facts of nature called spatial and temporal non-localities. In doing so we discover that we are dealing not merely with two types of non-locality but three. More accurately, the two complementary spatial and temporal non-localities "infer" the existence of a third type of non-locality whose presence cannot be directly confirmed.

Three Types of Non-localities

For convenience let us term spatial non-locality Type I non-locality, and temporal non-locality as Type II. Again, Types I and II "infer" the existence of Type III, which represents the entire physical situation. Although Types I and II taken together as complementary constructs describe the entire physical situation, neither can individually disclose this situation in any given instance. The reality represented by Type III non-locality is, in our view, the unified whole of space-time revealed in its complementary aspects as the unity of space (Type I non-locality) and the unity of time (Type II non-locality). Thus we can confirm with experiment the existence of Types I and II which taken together "imply" the existence of Type III, which is not itself subject to experimental confirmation.

Type III non-locality infers the existence, in short, of an undivided wholeness in the cosmos.[10] Spatial and temporal non-localities taken together mark the event horizon where we confront in the context of our knowledge of the physical situation the existence of the whole. Although spatial and temporal non-localities are present in acts of observation in astronomy, neither can be the subject of physical theory for the same reason that the results of the Aspect experiments proving Bell's inequality do not lead to additional theory. What each reveals is an aspect of reality as a whole rather than the behavior of parts which science coordinates with physical theory.

How, then, do these non-localities serve the progress of science? One obvious answer is that they serve to clarify how one should view the universe, and what can and cannot be known about the universe in terms of the new epistemological situation. But they could also serve science by suggesting that the future progress of scientific thought might lie in the direction of heretofore undiscovered complementary constructs. The epistemological situation suggests that it is observation that brings the complementary constructs of space-time into existence as features of our conscious efforts to coordinate experience with the parts that cannot be the whole. In conducting experiments we do not, therefore, "cause" the past to happen, or "create" non-local connections. We are simply making a demonstration of the existence of the part-whole complementarity in our efforts to coordinate our knowledge of the parts. What comes into existence as an object of knowledge was not, as we understand it, created or caused by us for the simple reason that it was always there—and the "it" in this instance is a universe which can be "inferred" to exist in an undivided wholeness.

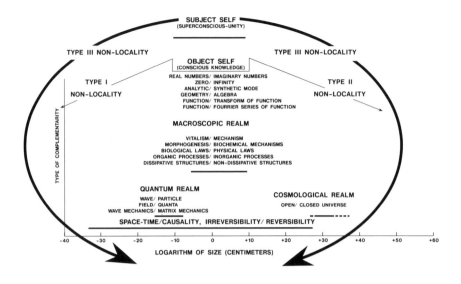

FIGURE 25. Complementarity universal diagram showing unfolding complementary relationships for different scales.

One interesting prospect here is that in some sense there may be no causal connections in reality-in-itself in that parts exist most fundamentally in this reality as the whole—they are, in other words, beyond space-time. The word "cause" in dealing with both spatial and temporal non-localities could merely be a term we apply in our conscious constructions of reality in the effort to coordinate tendencies to occur within the context of those constructions. Borrowing a phrase from Aristotle and refashioning it to suit this situation, the "unmoved mover" appears to be a rather appropriate description of the entire physical universe. All of which could mean for the future of science that we must learn to treat the universe as an undivided whole in the quantum mechanical sense, and that the principle of complementarity could be a guiding principle in uncovering increasingly more profound relations emergent from the whole. If this view is correct, then the task of science could be to frame empirical arguments to uncover that which will ultimately lie beyond empirical demonstration—the contentless background of the unified cosmos.

Complementarity and Cosmology

The suggestion that future progress in science may be marked by the discovery of additional profound complementarities can be reinforced with the realization that this is precisely the direction in which physics has been moving since 1905. Equally interesting, these complementarities seem to be emergent at increasingly larger times and scales. In order to illustrate both points, let us appeal to a type of diagram

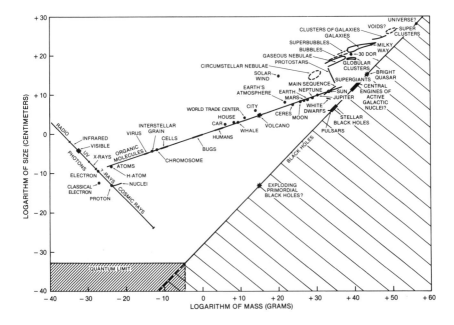

FIGURE 26. Mass versus size universal diagram for various classes of objects in the universe.

known as "universal diagrams."[11] The particular universal diagram shown in Figure 25 is based on all the known scales in the universe correlated with the evolving, or unfolding, order of the cosmos in terms of complementary relations. The horizontal axis represents scale, or size of objects in the universe, ranging some sixty-one orders of magnitude from the Planck length, 10^{-33} cm, to the radius of the observable universe, at about 10^{28} cm or roughly 10 billion light-years. The vertical axis roughly correlates emergent complementary relations with the scale at which they appear. In Figure 26 we show a universal diagram plotting the mass versus size of various classes of objects in the universe[13] which serves to provide the appropriate scales for the reader to be able to appreciate Figure 25.

At the beginning, or singularity, we witness the appearance of the first complementarity between the nothingness of the vacuum state and the somethingness of the quantum of action. It was, we now speculate, a fluctuation in the vacuum state caused by the quantum of action that resulted in creation. This first complementarity is then infolded into all subsequent events. Between 10^{-16} and 10^{-8} cm, the complementarities between field and quanta and particle-wave emerge, and are enfolded in the unified process. On roughly this same scale, there also emerges out of the unified field complementary relations between the four known fields interacting in terms of the enfolded complementarities of quantum of action-vacuum, quanta-field, and particle-wave. All of these complementarities become the basis for the part-whole complementarity that is evinced at all scales.

Although from the point of view of mathematical physics, a theory that allows us to understand the interaction of quanta and fields at all times and scales would appear to be a "theory of everything," there seems to be one aspect of the life of the cosmos this theory will not be able to describe—biological or organic life. Beginning at scales only a few thousand atoms across, or at a fraction of a micron, we confront the first viruses. These viruses, like all life forms, seem to display laws, or tendencies to occur, associated with life that cannot be fully comprehended by mathematical physics. In terms of scale, one could represent the realm of organic matter with the largest life-form known to us—the blue whale. But it is probably more appropriate, given the apparent interdependence and interconnectedness of all biological life, to represent it as the entire biosphere. Using the biosphere as the upper limit and the viruses as the lower limit, the scale for the emergence of the complementarities between organic and inorganic matter would range from a few microns to a few thousand kilometers. Note that all the additional and related complementarities we discussed earlier are also represented in the figure.

In order to complete this picture, let us now move into the realm of the astronomical beyond our solar system. Here we discover relatively small objects, like collapsed stars only a few kilometers across, as well as stellar black holes, neutron stars, terrestrial planets, collapsed white dwarfs, and much larger regular "main-sequence" stars. We also discover here red giant and red supergiant stars which can span dimensions larger than the inner solar system. Beyond the dimensions of all stellar objects fall the aggregates of stars, star clusters, and the gaseous nebular regions from which all stars are born—molecular regions, neutral hydrogen regions and ionized hydrogen regions. Moving into the galactic scale, we discover in ascending order tiny dwarf ellipticals with masses not much larger than the largest stellar globular clusters, the small irregular galaxies, the common spiral galaxies like the Milky Way, and, finally, the most massive and largest galaxies called the giant ellipticals. Outside of even this enormous scale we encounter giant new structures composed of clusters and superclusters of galaxies with dimensions spanning hundreds of millions of light years. As we saw earlier, the observable universe appears to be filled with these large superclusters.

As one approaches the inherent horizon of observability on these large scales, where we are dealing with fewer and fewer numbers of photons, the errors introduced in the observing process can, in our view, become so large due to the presence of quantum indeterminacy that the system under consideration can "bifurcate"[12] to either one of two relevant complementary constructs. This is analogous to the horizon of knowledge confronted in the quantum measurement problem where we can converge on either one or the other of the complementary aspects of the complete situation, like position or momentum, and yet can never measure or observe both aspects simultaneously. Although we can have an a priori "estimate" of the values of the two constructs within certain limits given by the Heisenberg relation, precise values cannot be predicted unless the system is forced into one of its characteristic states by an act of observation or measurement.

What we are suggesting is that if complementary relations are enfolded into the life of the universe throughout all scales, and become recognizable in mathematical physics as we approach the event horizon between part and whole, it could well be that *complementarity will also be useful in comprehending relations on the largest of scales and understanding the emergence of patterns at all levels characteristic of our universe.* Ultimately, we confront the event horizon of the whole as it has been extended thus far in time, namely, the entire universe. Our present candidate for an emergent complementarity on the largest scales is the open and closed universe constructs. [See Appendix]

The complementarities present in the mathematical theory (upper part of Figure 25) which allow us to coordinate such an impressive range of experience are obviously without scale. It was also, significantly, this inarguable fact which led to Cartesian dualism, and served to reinforce the hidden metaphysical presuppositions which are manifested in what we have termed the hidden ontology of classical epistemology. The discovery that these constructs are not pre-existing and disembodied forms has not been, as we have seen throughout, entirely welcomed by the mathematical physicist. If, however, we accept the conclusion that complementarities are fundamental to our conscious constructions of reality in both ordinary and mathematical language, and then factor in the conclusion that all events in the cosmos (past, present, and future) are enfolded into this seamless unity, the complementarities present in mathematical theory can be viewed in a quite different way.

From this perspective the description of physical reality in mathematical physics allows us to coordinate experience with this reality because the logical principle of complementarity is the fundamental dynamic in all our conscious constructions or representations of reality. Since the logical framework of complementarity, as we argued earlier, is more pristinely represented in mathematical language than it is in ordinary or everyday language, this could serve to explain why the language of mathematics, or the language of mathematical physics, is more "privileged" than ordinary language in the effort to coordinate experience with physical reality. Rather than assume that human consciousness is capable of uncovering the abstract mathematical and geometrical forms that are pre-existent in physical reality in the Cartesian sense, perhaps the more appropriate view is that our descriptions or representations of physical reality in physical theory serve to coordinate our experience with physical reality because the logical principle of complementarity is the fundamental structuring principle in this reality.

This hypothesis is made even more reasonable when we consider that complementarity appears to have manifested as a macro-level phenomenon in the evolution of life on earth beginning some 4 billion years ago. Since it is fair to say that the neuronal organization of the human brain is the most complex expression of the process of evolution to date, then the emergence of complementarity as a fundamental dynamic in the ability of the human brain to generate symbol systems, or conscious constructions of reality, can be viewed as a higher level manifestation of complementarity in the process of evolution. This does not, however, allow us to leap to the conclusion that there was any special "intent" or "design" at work in the evolution of the cosmos and/or in the evolution of life on earth that "culminates" in human consciousness. What we have here is merely a view of the evolution of the life of the cosmos which allows us to develop a more self-consistent and reasonable explanation for the emergence of human consciousness as an aspect of the life of this cosmos.

This argument also leads to an even more extraordinary conclusion. If consciousness is an emergent property of the universe in the case of human beings, would not this also imply, given the underlying wholeness of the cosmos, that the universe is itself conscious? In other words, if consciousness emerges out of this undivided wholeness at any stage in the evolution of the cosmos, would the temporal and spatial non-localities that are complementary aspects of this wholeness imply that it was present at all stages? We believe, providing of course that we refrain from viewing a conscious universe in anthropomorphic terms, that it does. Can we then leap to the conclusion that science has now proven that the universe is conscious? We cannot for the obvious reason that the quality of consciousness would be a property of reality-in-itself, or a property of the undivided wholeness which is "inferred" by our scientific description of the character of the universe but which is in no sense "disclosed" or "proven" to exist by that description. For these reasons, we seriously doubt that any future physical theory unifying general relativity with quantum theory will provide a simultaneous understanding of consciousness, as Roger Penrose advocates in his recent work *The Emperor's New Mind* [14].

Yet, since this wholeness appears to manifest itself in profound complementarities, and since we cannot in any single experimental situation or observation simultaneously affirm or confirm the existence of both complementarity aspects of the situation, the existence of this wholeness cannot be proven "in principle" by science. Since this represents an event horizon beyond which science cannot venture, the ultimate character of this wholeness, and even whether one chooses to recognize its existence or not, are now issues that lie, in our view, completely outside the domain of scientific knowledge. Put more simply, any scientific description of the parts, which manifest most profoundly in complementarities, can never be a description of the whole.

Let it be clear on this point lest this entire discussion be profoundly misunderstood. Science in our new situation cannot, in our view, say anything about reality-in-itself because the existence of the whole, which we have identified with reality-in-itself, can never be disclosed by mathematical physics. The simple and straightforward explanation as to why this is the case is that both aspects of the manner in which this whole manifests in physical reality cannot, in accordance with the logical principle of complementarity, be "simultaneously" disclosed. In what we view as the most dramatic demonstration of this truth, Types I and II non-locality cannot disclose or "prove" the existence of Type III non-locality, or reality-in-itself. If this is, as we have argued, a quantum mechanical universe, then the same conclusion must be drawn in all our attempts to coordinate our experience with nature. All we can do here, and this is all we have chosen to do, is to suggest that modern physical theory "infers" the existence of this wholeness in a manner that obviates any opportunity to experimentally confirm or "prove" its existence. If we assume that the existence of the whole has been "proven" by science, then we must also assume that repeatable scientific experiments under controlled conditions are not "essential" in the process of "validating" physical theory. Yet if we endorse this last assumption, we clearly, in our view, not only leave the domain of the truths of science but also, wittingly or unwittingly, obviate the epistemological authority of all such truths.

The fact that our new epistemological situation in science suggests that questions regarding the character of reality-in-itself no longer lie within the domain of science does not, however, oblige those of us who are concerned about the character of this reality, or our relationship to it, to conclude that science has rendered all such concerns meaningless. It simply means that such concerns no longer lie within the domain of science. If we choose, as we have done in the next chapter, to speculate upon what science "infers" about the character of reality-in-itself and our relationship to it, we are, of course, dealing in the realm of inference and, also, in our particular case, venturing into the realm of the metaphysical. To do so one must not only believe, as we believe, that metaphysical questions, particularly the ontological question, are valid concerns. One must also believe in the reality of $\mathcal{NO}$ Being. This reality, in our view, is most obviously affirmed on the intellectual level by the fact that the universe is something when it could just as well have been nothing, and on the personal, private, and subjective level by the "feeling" or "intuition" that we are profoundly connected to the whole.

If one shares this belief, and there are those, of course, who do not, then the fact that modern physical theory "infers" the existence of reality-in-itself without being able to disclose or "prove" the existence of this reality could obviously enliven the dialogue between members of C.P. Snow's two cultures. It is important to note here that the historical antipathy between these two ways of knowing can be largely explained in terms of the classical assumption that knowledge of all the constituent parts of the universe was equal to knowledge of the whole. Thus members of the two cultures have often in some sense been competing for definitions of reality-in-itself, or the ultimate nature of all Being. What seems clear at this point in time is that science must drop out of this competition—it cannot, in principle, provide a description of reality-in-itself. In the terms we used earlier, in our discussion of Heidegger, the sum of beings is not Being. On the other hand, humanists and social scientists are not likely to derive very much from a renewed dialogue with science if they ignore what science has to say about truth within the domain of science.

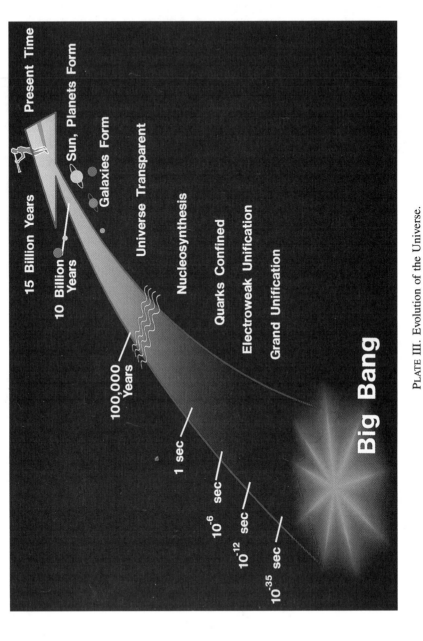

PLATE III. Evolution of the Universe.

9
The Ceremony of Innocence:
Science and Ethics

We do not know whether we shall succeed in once more expressing the spiritual form of our future communities in the old religious language. A rationalistic play with words and concepts is of little assistance here; the most important preconditions are honesty and directness. But since ethics is the basis for the communal life of men, and ethics can only be derived from that fundamental human attitude which I have called the spiritual pattern of the community, we must bend all our efforts to reuniting ourselves, along with the younger generation, in a common human outlook. I am convinced that we can succeed in this if again we can find the right balance between the two kinds of truth.[1]
—*Werner Heisenberg*

Although the world-view of classical physics allowed the physicist to presume that communion with the essences of physical reality via mathematical laws and associated theories was possible, it otherwise made no provisions for the knowing mind. Nature in classical physics was viewed as forces acting between mass points in the abstract background of space and time. Thus the universe became a vast machine in which collections of mass points interacted with one another in terms of external forces dependent only upon the masses and the effective dynamics between them. In this model the knowing self was separate, discrete, atomized, and achieved its knowledge of physical reality from the "outside" of physical systems, without disturbing the system under study.

As Henry Stapp puts it, "Classical physics not only fails to demand the mental, it fails to even provide a rational place for the mental. And if the mental is introduced ad hoc, then it must remain totally ineffectual, in absolute contradiction to our deepest experience."[2] In addition to providing a view of human beings as mere cogs in a giant machine linked to other parts of the machine in only the most mundane material terms, classical physics also effectively isolated the individual self from any creative aspect of nature. In the classical picture all the creativity associated with the cosmos was exhausted in the first instant of creation, and what transpires thereafter is utterly preordained.[3]

In our new situation, the status of the knowing mind is quite different. All of modern physics contributes to a view of the universe as an unbroken and undissectible dynamic whole. As Melic Capek points out, "There can hardly be a sharper contrast than that between the everlasting atoms of classical physics and the vanishing 'particles' of modern physics."[4] The classical notion of substance as composed of indestructible building blocks has been further undermined by the recognition that particles cannot be viewed as separate and discrete. As Stapp puts it,

...each atom turns out to be nothing but the potentialities in the behavior pattern of others. What we find, therefore, are not elementary space-time realities, but rather a web of relationships in which no part can stand alone; every part derives its meaning and existence only from its place within the whole.[5]

As Errol Harris notes, general relativity "identified field with space-time curvature," and showed that fundamental constants, such as the speed of light, physical laws and primary entities, such as mass, are "intimately linked with one another."[6] Any measurement in relativity theory is dependent upon the relative velocity of the reference frame in which the measurement is made, and every frame is related to every other frame. The relationship between moving frames can be visualized as a rotation of axes in the four-dimensional space-time continuum. It is the mutual relation of entities that determine their nature and properties as well as the curvature of space-time.

Similarly, in quantum physics the characteristics of particles and quanta are not isolatable given the interchangeability of the particle-wave complementary aspects, and their field interactions via exchanges of other quanta. Matter cannot be dissected from the omnipresent sea of energy, nor can we in theory or in fact observe matter from the "outside." As we voyage further into the realm of the unvisualizable in search of the grand unified theory incorporating all forces, the picture becomes increasingly holistic. The constituents of matter and the basic phenomena involved are unmistakably interconnected, interrelated, and inter-dependent. As Heisenberg put it, the cosmos "appears as a complicated tissue of events, in which connections of different kinds alternate or overlay or combine and thereby determine the texture of the whole."[7]

This means that a purely reductionist approach to a complete understanding of physical reality, which was the goal of classical physics, is no longer appropriate. The inability of the reductionist approach to completely comprehend or "subsume" physical reality with an appeal to physical theory is, as we noted in the introduction, one of the inescapable implications of Gödel's Incompleteness Theorem. Even though Kurt Gödel as a young mathematician was greatly influenced by the thinking of the Vienna Circle, his work "proved" that the principal aspiration of these theorists was "in principle" unattainable. Gödel's enormously important but often ignored theorem, which was developed in the summer of 1930, "proves" that mathematics, or the language of physical theory, cannot reach closure. Since no algorithm, or calculational procedure, that uses mathematical proofs can prove its own validity, any mathematical description which claims to have reached closure, or to have provided an exhaustively complete description of any aspect of physical reality, cannot prove itself. As the mathematician Rudy Rucker puts it:

> Mathematics is open-ended. There can never be a final best system of mathematics. Every axiom-system for mathematics will eventually run into certain simple problems that it cannot solve at all.[8]

If no mathematical system, no matter how formal, can reach closure, then it follows no physical theory built, as all modern physical theories must be, on mathematical systems, can reach closure. This means, in short, that even if we do construct a "Theory of Everything," and even if that theory could somehow coordinate within its mathematical framework an explanation of the phenomena of life and/or consciousness, this theory could not "in principle" claim to be a complete, final, or ultimate description.

The Incompleteness Theorem does not, however, imply that mathematical language and physical theories based on that language are not privileged in coordinating our experience with physical reality. As we have continually emphasized in this discussion, the requirement that a physical theory can be assumed valid only if its predictions hold in repeatable experiments under controlled conditions is fundamental to the "bounded" and "context drive" progress of scientific knowledge. The Incompleteness Theorem simply reveals that the language of mathematical physics, no matter what progress is made in the effort to better coordinate experience with physical reality, cannot "in principle" completely disclose this reality. What the theorem reveals in regard in the limits of mathematical language closely parallels in our view what is revealed in our new epistemological situation in a quantum mechanical universe—the universe as a whole, or reality-in-itself, cannot "in principle" be completely disclosed in physical theory. The general principle of complementarity, which applies to all scales and times, is, as we see it, a complement to the Incompleteness Theorem in that it provides the epistemological framework within which we can continue to better coordinate experience with physical reality with the clear understanding that reality-in-itself can never be finally disclosed or defined. Thus the ambition to develop a "Theory of Everything" must be viewed in our new situation as yet another unexamined legacy of the hidden ontology of classical epistemology. Simply put, the classical assumption that the collection of parts constitutes the whole has proven invalid. We now know that the properties of parts can only be understood in terms of the dynamics of the whole, and that what we call a "part" is a pattern in the inseparable web of relations. Equally important in this discussion, the unity and difference between parts and whole on the most fundamental levels inevitably discloses complementary relationships.

Part and Whole in Modern Physical Theory

As Harris notes in thinking about the special character of wholeness in modern physics, a "whole is always and necessarily a unity of and in difference."[9] The following discussion of the special character of wholeness in modern physical theory is based on Harris' lucid and detailed commentary of the subject. A unity without internal content is a blank or empty set, and is not recognizable as a whole. A collection of merely externally related parts does not constitute a whole in that the parts will not be "mutually adaptive and complementary to one another." Wholeness requires a complementary relationship between unity and difference, and is governed by a principle of organization determining the interrelationship between parts. This organizing principle must be universal to a genuine whole and implicit in all parts which constitute the whole, even though the whole is exemplified only in its parts. In other words this principle of order, as Harris notes, "is nothing real in and of itself. It is the way the parts are organized, and not another constituent additional to those that constitute the totality."[10]

In a genuine whole the relationships between the constituent parts must be *internal* or *immanent* in the parts, as opposed to a more spurious whole in which parts appear to disclose wholeness due to relationships that are *external* to the parts.[11] The collection of parts that would allegedly constitute the whole in classical physics is an example of such a spurious whole. Parts constitute a genuine whole when the universal principle of order is "inside" the parts, and thereby adjusts each to all so that they interlock and become mutually complementary. The terms and relations between parts in such a whole will have in common the immanent principle of order which is the "structure of the system to which they belong."[12] This is precisely the character of wholeness as has been revealed in both relativity theory and quantum mechanics.

The cosmos is a dynamic sea of energy manifesting itself in entangled quanta which results in a seamless wholeness on the most primary level, and in seamlessly interconnected events on any level. This is true from the contentless content of the singularity at the origins of the cosmos, to increasing levels of complementary relations in space and time, quantum and field, wave and particle, field and field, and atomic structure with atomic structure in the sharing of electrons as a result of the deeper order of quanta and fields. This dynamic progress eventually unfolds into a complementary relationship on the most complex of levels between organic molecules, which display biological regularities, and inorganic molecules, which do not.

Thus the unity that we witness in the description of physical reality in modern physics reveals a complementary relationship between the differences between parts which constitute content, and the universal ordering principle which is immanent in each of the parts and which cannot be finally disclosed in the analysis of the parts. It is our study of the differences between the parts that obliges us to uncover the dynamic structure of the whole present in each of these parts. Yet the part can never be finally isolated from the web of relationships which discloses the interconnections with the whole, and any attempt to do so results, as Bohr recognized, in ambiguity.

Yet order as we know it in the complementary relationship between difference and sameness in any physical event is never external to that event—the connections are immanent in the event. That non-locality would be added to this picture of the dynamic whole with Bell's theorem and the experiments testing that theorem is not really surprising—the relationship between part, as quantum event apparent in observation or measurement, and the undissectable whole, "inferred" but not "revealed" by the instantaneous correlations between measurements in space-like separated regions, is merely another logical extension of the part-whole complementarity evident throughout modern physics.

Each of the systems we attempt to isolate in the study of nature is in some sense a whole in that it represents the whole in the activity of being the part. But no single system, with the exception of the entire universe, can fully realize the cosmic order of the totality due to the partial and subordinate character of differentiated systems. No part can sustain itself in its own right for the reason that difference is only one complementary aspect of its being—the other aspect requires participation in the sameness of the cosmic order. All differentiated systems in nature require, in theory and in fact, supplementation by other systems.[13]

If we accept this view of higher, or more progressive, order in the universe, then each stage in the emergence of complementary relations allowing for more complex realizations of the totality "sublates" that which has gone before.[14] In other words, progressive order is a process in which every phase is a summation and transformation of the previous phase in that it incorporates the series of phases, or transformations, through which it has developed, and as a result of which it assumes its own form or structure. Process and end are now wedded in a difference/sameness complementarity with the organizing principle of the whole immanent throughout the proliferation of the constituent parts. This principle is immanent from the beginning, and discloses its potential, notes Harris, in the "entire scale of systematically (internally) related phases."[15] From this perspective, the best scientific account that we can ever achieve can only approximate the realization of the totality apparent to date.[16] And this is, incidentally, totally consistent with Gödel's Incompleteness Theorem. The ultimate organizing principle lies "in the outcome of the process and not its genetic origin." Reductionism is "ruled out," even presuming that quantum field theory will eventually provide a satisfactory account of the origins of the cosmos.[17]

Yet in discovering a new limit to our ability to fully comprehend physical reality, we are presented with a view of nature in which consciousness, or mind, can be properly defined as a phase in the process of the evolution of the cosmos implied in presupposing all other stages.[18] If it manifests or emerges in the later stages, and has been progressively unfolding from the beginning stages, then it would follow that the universe is in some sense conscious. Rather than merely view acts of cognition in classical terms as representations or images of independently existing facts, the empirical foundation for these acts in the physical substrate of the human brain must now be viewed as intimately connected with the whole. The dynamic neuronal patterns generated by the brain which are the basis for our "representations" of reality infold within themselves the previous stages of the life of the cosmos, and are seamlessly interconnected with all other activities in the cosmos. In the grand interplay of quanta and field in whatever stage of complexity, including the very activities of our brain, there is literally "no thing" that can be presumed isolated or discrete. In these terms we can "infer" that human consciousness "partakes" or "participates in" the conscious universe, and that the construct of the alienated mind, no matter how real feelings of alienation might be in psychological terms, is not in accord with the scientific facts.

David Bohm has made a similar argument in positing the existence of the "holomovement" which he associates with laws describing an implicit level of reality in an "implicate" order that is hidden below our experience with the "explicate order." In his view the laws of this alleged holomovement could be very similar to those of thought. "What happens in our consciousness," writes Bohm, "and what happens in nature are not fundamentally different in form. Therefore thought and matter have a great similarity of order."[19] Although it is "conceivable" that the laws Bohm envisions could be disclosed in new theory and eventually be subject to some form of experimental verification, the more likely prospect by far is that any such tendencies to reoccur are emergent properties of the interactions of all the parts and, therefore, not subject to disclosure in theory or experiment. The theory that Bohm now has in mind that could theoretically disclose some of the hidden dynamics of the implicate order would be a global hidden variable theory, as opposed to the local hidden variable theory which the experiments testing Bell's theorem appear to have made scientifically gratuitous. Our argument against the prospect of a global hidden variable theory is simply that the character of the whole as we have come to know it in modern physical theory suggests that global, or universal, variables would necessarily be emergent properties of the whole and, therefore, could not be disclosed in the behavior of any collectivity of parts no matter how large. If we assume the efficacy of any global hidden variable theory in the future, it would appear that we must do so in the absence of experimental verification which, as we saw earlier in the case of the three quantum ontologies, typically involves ontologizing some aspect of the physical reality. At the same time we are obviously very much in agreement with Bohm's conclusion that undivided wholeness implies that no categorical distinction can be made between the dynamics of human consciousness and those of the entire cosmos.

Since the suggestion that the universe is conscious is likely to be met by large measures of skepticism, if not outright ridicule, by some, perhaps we should pause for a moment and better explain what we mean by it. We do not mean that this consciousness is in any sense anthropomorphic, or that it embodies or reflects the conscious content of human consciousness. Our present understanding of nature neither supports nor refutes any conceptions of design, meaning, purpose, intent, or plan associated with any mytho-religious heritage. As Heisenberg put it, such words are "taken from the realm of human experience" and are "metaphors at best."[20] Although we can, of course, employ the scientific description of nature in an effort to legitimate such conceptions, this exercise is rather fruitless. The scientific description of nature is like a neutral screen upon which we can project a seemingly endless variety of these legitimations. And yet the projections inevitably seem illusory if we allow the totality of scientific facts more authority than our need to ascribe some human meaning to them. What we mean by conscious universe is in accord with the totality of those facts, and is anthropocentric only to the extent that it does answer to the very basic human need to feel that a profound spiritual awareness of unity with the whole cannot be deemed illusory from a scientific point of view.

Since the universe evinces on the most fundamental level an undivided wholeness, and since this wholeness in modern physical theory must be associated with a principle of cosmic order, or there would be no order, this whole manifests order in a self-reflective fashion. It must, in other words, be self-reflectively aware of itself as reality-in-itself to manifest the order that is the prior condition for all manifestations of being. Since consciousness in its most narrow formulation for human beings can be defined as self-reflective awareness founded upon a sense of internal consistency or order, we can safely argue that the universe is, in this sense, conscious. In order to avoid this conclusion, one must deny the existence of order as well as consciousness. Since the universe would not exist in the absence of the former, and physics would not exist in the absence of the latter, denying this conclusion based on appeal to scientific knowledge seems, in our view, rather futile.

Consciousness and the Single Significant Whole

The apprehension of the single significant whole as it is disclosed in physical theory and experiment would seem to be an indication that we have entered another and more advanced "stage" in the evolution of consciousness. What this theology of mind, or consciousness, assumes is the progressive realization of the totality of the organizing principle. This view becomes, we think, particularly compelling when we consider that complementary constructs appear to be as fundamental to our conscious constructions of reality in ordinary and mathematical languages as they are to the unfolding of progressive stages of complexity in physical reality. The suggestion is that human consciousness infolds within itself the fundamental logical principle of the conscious universe, and is thereby enabled to construct a view of this universe in physical theory which describes the unfolding of the cosmic order at previous stages in the life of the cosmos.

In order to take this argument a step further, let us return to the earlier distinction between the content of consciousness, in which we construct in both ordinary and mathematical language systems conceptions of reality, and the background of consciousness, in which we apprehend our existence prior to any conscious constructions. If consciousness, as we have defined it, is embedded in the universe, and if the evolution of consciousness progressively discloses in physics the totality of the universal principle of order implicit in all activities in nature, then the single significant whole whose existence is inferred, but not finally disclosed, in the conscious content of physical theory can be "assumed" to be ontologically grounded in the life of nature. The most dramatic example of this situation is, in our view, the three types of non-localities discussed in the last chapter. Although we can confirm the existence of both Type I (spatial) and Type II (temporal), these complementary constructs can only "infer" the existence of the undivided wholeness represented by Type III. Although the two experimentally confirmable non-localities may bring us to the horizon of knowledge where we confront the existence of the undivided whole, we cannot cross that horizon in terms of the content of consciousness. Yet the fact that we cannot disclose this undivided wholeness in our conscious constructions of this reality as parts does not mean that science invalidates the prospect that we can apprehend this wholeness on a level that is prior to the conscious constructs. It merely means that science qua science cannot fully disclose or describe the whole.

No!

What this situation allows, which classical physics could not, is a rational place for the mental, and thereby demolishes the notion that the realm of human thought and feeling is merely ad hoc. Life and cognition in these terms can be assumed to be "grounded" in the single significant whole. Yet this is not a conclusion that can be "proven" in scientific terms for all the reasons discussed earlier. Since this single significant whole, or Being, must be represented in the conscious content as parts, or beings, it is not and cannot be a direct object of scientific inquiry or knowledge. Thus any direct experience we have of this whole is necessarily in the background of consciousness, and must be devoid of conscious content.

Herein lies the paradox. Scientific knowledge conditions us into an awareness of the existence of this whole, and yet cannot fully affirm or "prove" its existence in strictly scientific terms. Yet this would also argue that all those who apprehend the single significant whole, or experience cosmic religious feeling, with or without the awareness of the existence of the principle of cosmic order, are engaged in similar acts of communion with the whole. Yet any translation into conscious content of that experience, in scientific or religious thought, invokes reductionism where it cannot be applied. Put differently, all knowledge in the conscious content is a differentiated system that cannot by definition articulate the universal principle of order. Just as there can be no one-to-one correspondence between physical theory and physical reality, there can be no such correspondence between religious descriptions of beings and Being itself.

The best religious thinkers in virtually any religious heritage that one can name have, in general, been inclined to accept this inherent limitation in religious thought. Yet the community of physicists has not, in general, been inclined to accept such a limitation in their knowledge field for reasons that we are now familiar with. Classical physics featured a metaphysics which fostered the assumption that a description of the parts would theoretically disclose the workings of the deterministic whole. It is the emotional and/or psychological attachment of physicists to this metaphysics which translates into an apparent refusal to accept the proposition that knowledge of the parts does not culminate in disclosure or knowledge of the whole.

This attitude also serves to explain why science could easily appear as hostile to religion. A science that promises to subsume the universe within a comprehensive causal explanation for all physical events from the very beginning of the cosmos is an obvious threat to religion because the object of religion is the ineffable and the mysterious. In the absence, wrote Karl Jaspers, of any "sense of the infinite vastness that lies beyond our grasp, all we succeed in conveying is misery—not tragedy."[21] The evidence for the existence of the ineffable and mysterious disclosed by modern physics is as near as the dance of particles that make up our bodies, and as far as the furthest regions of the cosmos. The results of the experiments testing Bell's theorem suggest that all the parts, or any manifestation of "being" in the vast cosmos, are seamlessly interconnected in the unity of "Being." Yet quantum physics also says that the ground of Being for all this being will never be completely subsumed by rational understanding.

A New Basis for Dialogue Between Science and Religion

What all of this is meant to suggest is that we now have a new basis for meaningful dialogue between science and religion. Yet those who have the capacity for profound religious impulse who enter this dialogue in the effort to evolve a spiritual pattern for a global community will obviously have to overcome a rather large obstacle. Although scientific knowledge can no longer be viewed as obviating in any sense the most profound religious impulse—the apprehension of the single significant whole—that fact is likely to be lost on those who insist that scientific truths must legitimate anthropomorphic versions of religious truths. The assumption that one must make to enter this dialogue, and it may be quite impossible to make for many true believers, is that Being assumes the anthropomorphic guise of our particular conceptions of beings in a cultural context. Religious truth, like scientific truth, must be viewed as metaphor for that which we cannot fully describe.

Another of the large obstacles in the way of a renewed and meaningful dialogue between science and religion, as Kant was among the first to point out, is that we use two kinds of reason in coming to terms with reality—"theoretical reason" and "practical reason." Although theoretical reason in science has disclosed that the life of the cosmos is marvelously whole or unified in accordance with the principle of cosmic order, practical reason, which we employ to coordinate experience in everyday life, still obliges us to act "as if" we are discrete entities moving through the separate dimensions of space and time. Another large related problem is that practical reason is accrued in any culture in accordance with efforts to coordinate experience with macro-level visualizable phenomena, and thus tends to be more in accord with classical assumptions about the character of physical reality.

And yet it is also demonstrably true that theoretical reason does over time refashion the terms of construction of human reality within particular linguistic and cultural contexts, and thereby alters the dynamics of practical reason. As many scholars have exhaustively demonstrated, the classical paradigm in physics has greatly influenced and conditioned our understanding and management of human systems in economic and political reality. Virtually all models of this reality treat human systems as if they consisted of atomized units which interact with one another in terms of laws or forces external to the units. These laws or forces are also assumed to act upon the isolated or isolatable units to form hierarchical organizations which are themselves isolated or isolatable from other such organizations. The same laws or forces are then treated as acting externally upon the isolated or isolatable hierarchical systems to form hierarchical megasystems, and the sum of these parts is taken to be the whole. If the laws or forces, like those of capitalist or Marxist economic theory, are reified, then the growth and expansion of competing systems is assumed to move toward closure, or a situation in which all units that can be acted upon by these laws or forces are completely subsumed by them in one unified system. For this reason the true believers in the "real" existence of these economic laws tend to assume that progression toward closure is quite inevitable.

If theoretical reason in modern physics does eventually refashion the terms of constructing our symbolic universe to the extent that it impacts practical reason, then conceiving of a human being, as Einstein put it, as "part of the whole" is the leap of faith that would prove most critical. It is only in making this leap that we can begin, as he suggests, to free ourselves of the "optical illusions" of our present conception of self as a "part limited in space and time," and to widen "our circle of compassion to embrace all living creatures and the whole of nature in its beauty."[22] Yet one cannot, of course, merely reason or argue oneself into an acceptance of this proposition. One must also have the capacity, in our view, for what Einstein termed "cosmic religious feeling." Hopefully many of those who have the capacity will also be able to communicate their awareness to others in metaphoric representations in ordinary language with enormous emotional appeal. The task that lies before the poets of this new reality has been nicely described by Jonas Salk:

> Man has come to the threshold of a state of consciousness, regarding his nature and his relationship to the Cosmos, in terms that reflect "reality." By using the processes of Nature as metaphor, to describe the forces by which it operates upon and within Man, we come as close to describing "reality" as we can within the limits of our comprehension. Men will be very uneven in their capacity for such understanding, which, naturally, differs for different ages and cultures, and develops and changes over the course of time. For these reasons it will always be necessary to use metaphor and myth to provide "comprehensible" guides to living. In this way, Man's imagination and intellect play vital roles in his survival and evolution.[23]

The Emergence of the Ecological Paradigm

Granted that the probability for success in this formidable enterprise does not at the moment appear very great. Yet it also seems reasonable to assume that any hopes that we have for the continued survival of the species could rest rather squarely on the shoulders of those who are capable in the dialogue between science and religion of generating metaphors and myths that provide comprehensible guides to living. There are many who believe that the implications of the vision of reality contained in modern physical theory are already occasioning a massive restructuring of the terms of constructing our symbolic universe. The phrase that is most often used to describe this alleged revolution in thought is "paradigm shift." Since "paradigm" has taken on many diverse meanings, the word as it is used here is defined as follows: A paradigm is a constellation of values, beliefs, practices, and perceptions shared by a community, which governs the manner in which the community organizes itself.

Recently Fritjof Capra has attempted to make the case that one of the indications that this paradigm shift is in progress is the emergence of what he terms an "ecological world view."[24] On the most fundamental level, says Capra, ecological awareness is a deeply religious awareness in which the individual feels connected with the whole, as in the original root meaning of the word religion from the Latin "religare"—"to bind strongly." The ecological world-view, or social paradigm, is distinguishable, he suggests, in terms of five related "shifts" in emphasis which are entirely consistent with the understanding of physical reality revealed in modern physics. The five shifts are briefly summarized as follows:

1. *Shift from the Part to the Whole*—The properties of the parts must be understood as dynamics of the whole;

2. *Shift from Structure to Process*—Every structure is seen as the manifestation of an underlying process, and the entire web of relationships is understood to be fundamentally dynamic;

3. *Shift from Objective to "Epistemic" Science*—Descriptions are no longer viewed as objective and independent of the human observer and the process of knowledge. Thus this process, which we have defined as the new epistemology, must be included explicitly in the description;

4. *Shift from "Building" to "Network" as Metaphor of Knowledge*—Since phenomena exist by virtue of their mutually consistent relationships, any physics which describes phenomena must meet the requirement that components be consistent with one another and with themselves. Thus knowledge can no longer be viewed as "built" upon unchanging or "reified" foundations, and must be viewed rather as an interconnected network of relationships founded on self-consistency and general agreement with facts;

5. *Shift from Truth to Approximate Descriptions*—Since nature is an interconnected, dynamic web of relationships, the identification of patterns as objects depends upon the process of knowledge and human observation. This means that the true description of any object is a web of relationships associated with concepts and models, and that the whole which constitutes the entire web of relationships cannot be represented in this necessarily approximate description.

Although one could extend and refine this list, these shifts, or new terms for the construction of human knowledge, follow closely upon understanding our new situation in physics. They are not arbitrary in the least. What this suggests is that if thoughtful people are enticed into this understanding, and if they elect to re-examine the character of human knowledge and belief in terms of this model, they will draw remarkably similar conclusions. Yet Capra, as we see it at least, is quite correct in suggesting that this understanding will not lead to substantive changes in attitude and behavior unless intellectual understanding of the character of physical reality is wedded to profound religious or spiritual awareness. If those who have the capacity for this awareness are to become the poets and philosophers of this new reality, they will obviously require something more than a passing acquaintance with the character of physical reality as we know it in modern physics. This suggests that the study of science is vital to the future of man not merely because it provides greater levels of mastery and control over his environment or the basis for building new technologies. It could, in fact, play a vital and central role in developing that spiritual pattern for the entire global community which now appears to be our best hope for coordinating experience in the interest of survival and allowing human consciousness to evolve into higher and more progressive stages of awareness.

Systems Theory

We do not mean to suggest here, however, that those who do not, for whatever reasons, have any interest in ontology and/or feel that the vision of physical reality disclosed in modern physical theory has nothing to do with ontology cannot or should not appeal to the implications of modern physical theories in arriving at an improved understanding of better ways to coordinate human experience in the interest of survival. This is anything but the case, and the emergence and rapid growth of "systems theory," which is not normally concerned with metaphysical or ontological questions, provides eloquent testimony of this fact.

The Austrian biologist Ludwig von Bertalanffy is generally recognized as the father of systems theory. Drawing upon his studies in biology in the early 1920s, Bertalanffy was among the first to conclude that a mechanistic approach to understanding living systems could not comprehend or explain the essential dynamics of these systems. In the 1930s and 1940s, he attempted to demonstrate that abstract mathematical models could help scientists better comprehend the "problems of order, organization, wholeness, [and] theology" which were not comprehensible in the mechanistic approach.[25] The central or primary ambition here was to create the foundations for a kind of metadiscourse about wholes and their relations that could allow students in both the natural and human sciences to perceive isomorphisms in their work and better communicate with one another. Eventually numerous students in disciplines as diverse as psychology, sociology, economics, anthropology, and biology drew upon and reconceptualized Bertalanffy's terms and models with the result that the bibliography of systems theory studies has become quite massive.

According to Anthony Wilden in *Systems and Structure*, the principle sources of systems theory, several of which were recognized by Bertalanffy himself, were non-Euclidean mathematics, the gestalt psychology of Wolfgang Kohler, Levi-Strauss's anthropology, information theory and cybernetics, and quantum physics.[26] Wilden also attempts to characterize the paradigm shift occasioned by systems theory in terms of new epistemological "redirections" in the following manner:

FROM	TO
stasis	process
entity	relationship
atom	gestalt
aggregate	whole
heap	structure
part	system
analytics	dialectics
causality	constraint
energy	information
bioenergics	communication
equilibrium theory	negative entropy[27]

Systems theory regards living systems as "wholes" in terms of dynamic processes that combine energy and information in reciprocal relationships. Since living systems are always open and interactive with other systems, for reasons we demonstrated earlier, they cannot be described in classical terms, or in terms of linear chains of cause and effect. And this also means, obviously, that the atomistic and reductionist approach to living systems simply cannot work. If we are to examine differences in living systems, the emphasis in systems theory is on differentiations in structure and form that correlate with different modes of carrying and exchanging information. The privileged model of living systems for systems theorists is the ecosystem which is presumed to exist as a whole, and which cannot be studied in the absence of the assumption that the observer and the observed system are part of this whole. Many systems theorists also recognize that either-or logic is inadequate to analyze the both/and relations between living systems, and that such relations must be viewed as complementary. That this understanding of living systems as well as the epistemological redirections taken in systems theory listed by Wilden are wholly commensurate with our understanding of the dynamics of physical reality in a quantum mechanical universe should, at this point, be obvious.

One of the most exciting and interesting conclusions that have recently been drawn by systems theorists has to do with a view of human beings as living systems within the context of the whole, defined in this instance as the ecosystem. As Susan Stewart suggests in *On Longing*, human consciousness can be viewed as a "feedback" system comparable to the feedback system of lower organisms. Since human beings have historically used their bodies as the measure of scale, she also argues that our prior sense of personal scale will be radically transformed when the size and number of our relations to other feedback systems in the ecosystem are recognized and understood.[28] Michael Serres in *The Parasite* describes his own body in terms of such scales in the following manner: "My body is an exchanger of time. It is filled with signals, noises, messages, and parasites. And it is not at all exceptional in this vast world. It is true of animals and plants, of air crystals, of cells and atoms, of groups and constructed objects. Transformation, deformation of information."[29] What is particularly interesting here from our perspective is that a similar view of our bodies in relation to the activities of the entire cosmos on any scale is easily derived from an understanding of the relation between parts and whole in a quantum mechanical universe in which non-locality is a new fact of nature.

What is most intriguing here for the purposes of this discussion are the rather stunning number of parallels between the paradigm of systems theory and the paradigm associated with the organization and dynamics of physical reality provided by modern physical theory. The existence of these parallels serves, as we see it, to reinforce the view advanced earlier that human consciousness has entered, or is in the process of entering, a more advanced stage in the evolution of consciousness. And yet as the success of systems theory suggests, one can make effective use of that paradigm in the effort to better coordinate human experience in the interests of survival without being required to venture into the realm of the metaphysical or the ontological, and the same applies to the paradigm provided by modern physics. And there is also, as we have repeatedly emphasized, nothing in the vision of physical reality provided by modern physical theory that requires one to endorse any metaphysical or ontological positions.

It is, of course, conceivable, that the human community will learn to effectively coordinate its experience in the interests of survival based on a pragmatic acceptance of an improved understanding of the actual conditions and terms for survival. We are, however, personally in agreement with Capra who has rather consistently argued that a belief in ontology, or in the existence of a Being that is not and cannot be the sum of beings, will be a vital aspect of the global revolution in thought that now seems to be a prerequisite for the survival of our species. In practical or operational terms, this must, in our view, be the case because the "timely" adjustments that we must make in the efforts to better coordinate experience in the interests of survival will require great personal sacrifice, particularly on the part of members of economically privileged cultures. It also seems clear to us, as it does to Capra, that a willingness to sacrifice oneself for the good of the other, or for the good of the "whole," has rarely occurred in the course of human history as the direct result of a pragmatic intellectual understanding of the "necessity" to make such sacrifices.

Sacrifice on this order requires a profound sense of identification with the "other" that operates at the deepest levels of our emotional lives. And it is this sense of identification which has, quite obviously, always been one of the primary challenges and goals of religious thought and practice. Yet one clearly does not arrive at a belief in an ontology, which has typically been the only way in which human beings have achieved a truly profound sense of identification with the "other," based on the "practical" necessity of doing so. Such a belief requires, as Kierkegaard pointed out, a "leap of faith" that may have little or nothing to do with the dictates of reason.

What makes us confident that ontology will play a primary role in the global revolution in thought which now seems to be in process is simply the "conviction" that the vast majority of human beings are willing and able to make such a leap, particularly when the modern scientific world-view gives them a new "freedom" to do so. In our view, the majority of human beings do apprehend or intuit on the deepest levels of their subjective experience the existence of Being as a self-evident truth. What is, therefore, required, as we see it, to transform that which seems self-evident into the basis for a profound and active sense of identification with the other that could serve as the basis for sacrificing self for the other, or for the good of the whole, is a much improved dialogue between the truths of science and those of religion. Yet it also seems clear that this dialogue will not occur on the scale that we are imagining here in the absence of some dramatic improvements in the manner in which we "teach" science.

A New Program for Scientific Education

The study of science we have in mind is not the pallid and dispirited version that is frequently taught in public schools and, often enough, in college or university classrooms. It is a science self-reflectively aware of its origins, its transformations, and its inherent limits. This science is not a meaningless number crunching and equation solving activity. It continuously emphasizes the essential wonder and beauty of itself and the universe it describes, is aware of its creators and their dreams and struggles, and consistently advertises, as the best scientists have always done, that it is the most communal of all our ways of knowing. Most important, it remains aware in the analysis of "parts" of the greatest and most startling of scientific facts—the ceaseless interconnectedness and interpenetration of the whole. Although this science will never compromise its truths out of deference to the anthropomorphic truths of religion, it is also aware that it should not, or rather cannot, legislate over the character of the most profound religious truth because ontological questions no longer lie within the province of science.

Religion, in turn, could enter the dialogue with the recognition that science neither wants nor needs to challenge its authority— the ontological question no longer lies within the domain of science. If religion elects to challenge what science does know within its own domain, it must either withdraw from the dialogue or engage science on its own terms. Applying metaphysics where there is no metaphysics, or attempting to rewrite or rework scientific truths and/or facts in the effort to prove metaphysical assumptions, merely displays a profound misunderstanding of science, and an apparent unwillingness to recognize its successes. But since the study of science could serve profoundly religious truths while not claiming to legislate over the ultimate character of these truths, principally that the cosmos is something as opposed to nothing and that this something is a single significant whole, religion in its own interests should become an avid promoter of the study of science.

What we are trying suggest here is that if the dialogue between the truths of science and religion were as open and honest as it could and should be, then we might begin to discover a spiritual pattern that could function as the basis for a global human ethos. Central to this vision would be a cosmos rippling with tension

⊗ Science is a true "religious truth seeker".

evolving out of itself endless examples of the awe and wonder of its seamlessly interconnected life. And central to the cultivation and practice of the spiritual pattern of the community would be a profound acceptance of the astonishing fact of our being. Equally important, this vision also requires that we appreciate that the conditions for our being and becoming were conducive to a radical freedom—the freedom which allowed us to fashion our own reality.

If the dross of anthropomorphism can be eliminated in a renewed dialogue between science and religion, the era in which we were obliged to conceive, as Koyré put it, of the truths of each way of knowing as two truths and, therefore, as providing no truth at all, could be over. Science in our new situation in no way argues against the existence of God, or Being, and profoundly augments the sense of the cosmos as a single significant whole. That the ultimate no longer appears to be clothed in the arbitrarily derived terms of our previous understanding simply means that the mystery that evades all human understanding remains. The study of physical reality should only take us perpetually closer to that horizon of knowledge where the sum of beings is not and cannot be Being, while never being able to comprehend or explain this mystery. As William Blake suggested in the age of Newton, the "bounded is loathed by its possessor," and what loathing we would surely feel if we had discovered that the meaning of meaning was only ourselves.

Wolfgang Pauli, who also thought long and hard about the ethical good that could be occasioned by a renewed dialogue between science and religion, made the following optimistic forecast:

> Contrary to the strict division of the activity of the human spirit into separate departments—a division prevailing since the nineteenth century—I consider the ambition of overcoming opposites, including also a synthesis embracing both rational understanding and the mystical experience of unity, to be the mythos, spoken and unspoken, of our present day and age.[30]

And Pauli also suggested, as we have, that the logical framework of complementarity could serve in helping us to create such an ethos.

This is a project which will demand a strong sense of intellectual community, a large capacity for spiritual awareness, a profound commitment to the proposition that knowledge coordinates experience in the interest of survival, and an unwavering belief that we are free both to elect the best means of our survival and to develop our humanity to its fullest potential. The essential truth revealed by science which the religious imagination should now begin to explore with the intent of enhancing its ethical dimensions was described by Schrödinger as follows:

> Hence this life of yours which you are living is not merely a piece of the entire existence, but is, in a certain sense, the whole; only this whole is not so constituted that it can be surveyed in one single glance.[31]

Virtually all major religious traditions have at some point featured this understanding in their mystical traditions in constructs like the Nun of the Egyptians, the En-Sof of the Hebrews, the Father of the Judeo-Christians, the Allah of the Sufis, the Tao of the Chinese, the Brahman of the Vedantins, and the Paramashiva of the Shaivites.[32] And the history of religious thought reveals a progression in virtually all religious traditions toward the conception of spiritual reality as a unified essence in which the self is manifested, or mirrored, in intimate connection with the whole. This would seem to suggest that the evolution of human consciousness in both scientific and religious thought is toward the affirmation of the existence of the single significant whole, and that these two versions of the ultimate truth exist in complementary relation. It is also important to note that the scientific world-view, as Schrödinger also appreciated, simply cannot in itself satisfy our need to better understand the character of ultimate truth:

> The scientific picture of the real world around me is very deficient. It gives me a lot of factual information, puts all our experience in a magnificently consistent order, but it is ghastly silent about all and sundry that is really dear to our heart, that really matters to us.[33]

It is time, we suggest, for the religious imagination and the religious experience to engage the complementary truths of science in filling that silence with meaning.

Appendix

Horizons of Knowledge in Big-Bang Cosmological Models

In any big-bang model of the universe we encounter horizons of knowledge which arise because a particular model has been adopted. The most obvious horizon has to do with the fact that in any big-bang model of the universe, the universe becomes opaque to its own radiation for sufficiently early times. The best way to describe the situation is in terms of the so-called redshift. We can express the age of the universe as a function of the fractional shift in wavelength, or "redshift" z, where z is defined as the ratio between the change of the wavelength, $\lambda_o - \lambda_e$, over the emitted wavelength λ_e and λ_o is the observed wavelength seen by a moving observer. One hundred thousand years after the beginning, when the universe was only 0.1% of 1% of its present age, corresponds then to z of 1000.

The most distant quasars, in contrast, are seen at a redshift of slightly above 4, and emitted their observable light when the universe was about 10% of its present age. Although radiation can in principle tell us much more about the early universe than matter, the opaqueness of the universe prior to z = 1000 simply does not allow us to trace or confirm the origin of the universe in the big-bang cosmology using photons. Yet it is photons that provide virtually all information about both the large-scale structure of the universe and what we know of its evolution. It is, therefore, at z = 1000 that we encounter the first horizon of knowledge about our universe in any big-bang theoretical model. If our only access to reliable information about the earlier universe must be based on the observation of photons, that horizon is, *in principle*, impregnable.[1]

The hope has been that we can discover other means of uncovering clear-cut evidence about the early universe based on information provided by quanta other than light. Yet it is unlikely that any other means will provide evidence as clear-cut as that of light about the universe we live in. Scientists were recently teased with the suggestion that other quanta could provide such evidence as their measuring instruments recorded an unanticipated shower of neutrinos. Emitted by the exploding supernova in the Large Magellanic Cloud and traveling at the speed of light for more than 150,000 years, these neutrinos reached the earth on February 22, 1987. The primary reason the scientific community was excited by this event is that it provided important evidence about the mechanism of supernova explosion. But there was also hope that the neutrinos would provide information with cosmological import. Some particle physicists had postulated that neutrinos possess a rest mass with a value which, although small, is not exactly equal to zero and, therefore, that all the neutrinos in the universe together could provide enough "missing mass" to close the universe. Although only 17 neutrinos from the supernova were observed from the supernova by teams in the U.S. and Japan, the evidence nevertheless suggested that the neutrinos appear to have a rest mass exactly equal to zero, and, if they do have some rest mass, it is not appreciable enough to close the universe.

But this does not mean that neutrinos cannot, in principle, provide important information about the early universe. If primordial neutrinos emitted a few seconds after the beginning of the universe, or at a redshift $z \sim 10^9$, were ever observed, that would be an exciting event. Although it is anticipated that their numbers would be much greater than those produced by stellar events such as supernova explosions, their energies would be billions of times less. Ignoring for the time being the great difficulty in detecting primordial neutrinos, successful observation of these neutrinos would still not yield sufficient information about the very early universe to answer fundamental questions. Even the very high redshift $z = 10^9$, at which primordial neutrinos would have been emitted, is not sufficiently large to allow us to probe timescales near the big bang itself or near the hypothetical inflationary period at 10^{-35}. This means that the observational horizon of knowledge at $z \sim 10^9$ is the ultimate horizon from which we can access direct information about the universe.[2]

The other major source of information about the early life of the universe comes from experiments in particle accelerators. Although studies of ordinary matter in high energy physics can yield a lot of detailed information about the hypothesis of element formation in the early universe,[3] the actual situation is not so rosy. The uncertainties in the abundances of the primordial elements, like deuterium, regular helium, and ^{7}Li, are large in the big-bang models. Also, the details of big-bang models are least sensitive to the abundance of regular helium, which is by far the most abundant of primordial elements formed from hydrogen. Present results, taken at face value, imply a mean density of baryons about 3×10^{-31} gr/cm^3, about two orders of magnitude less than the critical density required to close the universe. What is most significant here is that the results imply an open universe. Space missions, like the Hubble Space Telescope, will also search for better estimates of the abundance of the primordial elements. Yet even if we could somehow identify in the future what part of these elements was primordial, or synthesized in the first three minutes, we could still not go further back in redshift than $z = 10^8$.

Of all the above mentioned horizons of knowledge at $z = 10^9$, 10^8, and 1000, it is the last that is likely to remain the only one we can explore for the foreseeable future. The neutrino horizon at 10^9 is not going to be qualitatively different from the photon horizon. Since at early periods both photons and neutrinos were in thermal equilibrium with matter, both primordial neutrinos and photons follow a black body distribution because both were in equilibrium with themselves and everything else early on. Thus, whatever problems of interpretation we are facing today with regards to the background photons will not go away even if we do manage to observe primordial neutrinos. We should also not anticipate that high energy physics will probe the element horizon prior to the first three minutes. The problem here is that we cannot simulate the complexity of nuclear reactions applicable at that time in the early universe due to the inherent uncertainties of the reactions themselves.

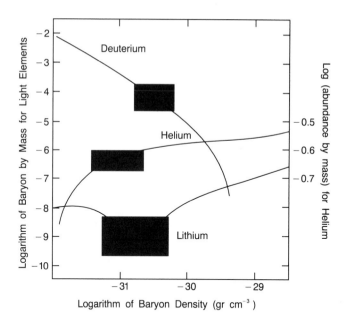

FIGURE 27. Abundances of primordial light elements. Taken at face value, these results imply an open universe with density of matter in the range $10^{-31} - 10^{-30}$ gr cm $^{-3}$.

The last observational problem in big-bang cosmology we want to mention involves the measurements of redshifts of the light spectra of distant galaxies. In the effort to test the geometry of the universe, one studies the Hubble Diagram, or a diagram of magnitude versus redshift using "standard candles", i.e., distant galaxies which supposedly have a luminosity that does not vary with time. One would then be able to determine the curvature effects of the universe in the Hubble Diagram. Although quasars can be seen as far away as z = 4 or more, they are notoriously unreliable as standard candles because of the tremendous range of luminosities they present for any redshift. The traditional approach has been to search for other standard candles. It would seem that the best choice would be ordinary galaxies, like the Milky Way. But most ordinary galaxies cannot be observed from the ground beyond a redshift of about unity and will only begin to be observable at redshifts of four or more when the Space Telescope starts looking at their spectra in 1990. In the past, astronomers have relied on bright radio galaxies as standard candles. These are seen much further away than ordinary galaxies. The most distant galaxy discovered so far has a redshift of 3.8 at a corresponding distance of 15 billion light-years.[4] Unfortunately radio galaxies have turned out not to be "standard" after all;[5] their light changes significantly with distance and/or distant radio galaxies are different from nearby ones. It remains to be seen whether distant ordinary galaxies will turn out to be "ordinary" after all. It may turn out that all galaxies look sufficiently different when the universe was 10% of its age

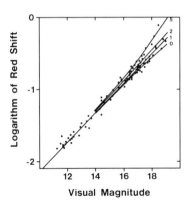

FIGURE 28. The Hubble Diagram and the various associated geometries for distant galaxies. The curves are labelled according to a parameter: 0–1/2 (open); 1/2 (Einstein-de Sitter); > 1/2 (closed) (Data from M. Rowan-Robinson, *The Cosmological Distance Ladder*, New York: W.H. Freeman, 1985).

compared to today. Redshifts in the range 5–10 are particularly important to study since astronomers suspect that it is in this range where we would detect light from galaxies that began to form more than 15 billion years ago. Also, one requires redshifts of this magnitude to distinguish which big-bang model is applicable when one studies the Hubble Diagram (Figure 28) because the observational uncertainties are so large that one cannot distinguish between competing models for smaller redshifts. Assuming that the quest for "standard candles" is successful one encounters a difficulty which may turn out to be insurmountable. Unfortunately, images of ordinary galaxies begin to merge at those redshifts.[6] This can be easily shown if one examines how the apparent size of a finite-size object varies with distance in the big-bang cosmology. It is known that the apparent size of an object of some "standard" length (say a galaxy of 60,000 light-years across) subtends in the sky decreases as the distance increases. General relativity predicts that above a certain distance, which in most big-bang models corresponds to a redshift of about one, the apparent size will then continue to increase as the distance increases. This happens because of curvature effects. At some redshift, then, which turns out to be greater than 1 but less than 10, in most cases about 5, the apparent size of a distant galaxy tens of thousands of light-years across would be a few arc seconds, and would then become comparable to the distance between itself and other neighboring galaxies, also a few arc seconds. This statement, which has not been appreciated in cosmological theory, is based on extrapolating the observed distribution of faint galaxies in the sky to fainter and fainter magnitudes, and the exact value of the redshift at which it occurs is subject to uncertainties about the distant and presently unobserved ordinary galaxies. Nevertheless, the overall argument remains valid. If this is born out by future space-borne observations, it will be observationally impossible to obtain an accurate spectrum of any galaxy beyond that redshift, and it will then be impossible to test which of the competing big-bang models is applicable.

The geometry of the universe and the appropriate applicable model will, therefore, be indeterminate. This "galaxy image horizon" is much too close to us at fairly low redshifts and certainly not much exceeding 10. We would then suggest that the experimental methods in use since the time of Hubble which have led to the hypothesis of the expanding universe will probably not be useful in better verifying that hypothesis. Spectra of different sources in the sky would themselves be blended together as one looks at fainter and fainter sources since the images would begin to merge. Eventually, the background from different galaxies would dominate the spectrum from a single distant galaxy, and reliable spectra could not be obtained to test the geometry of the universe.

Hypothesis of Large-Scale Complementarities

In our view, the reason the universe is observed to be so close to the flat, or Euclidean, model is that this model represents the division between the complementary constructs of open and closed cosmologies. Rather than attempting to understand the observed flatness in terms of a single construct, the introduction of complementarity to cosmology allows seemingly disparate constructs to be brought together into the picture. As our observations carry us to deeper and deeper regions of space, the inherent uncertainties of specific big-bang models (e.g., the question when galaxies formed or how distant galaxies differ from nearby galaxies) will not allow us to unequivocally prove the correctness of one of these cosmological models. To put it differently, the observational errors which arise from the experimental situation coupled with the inherent theoretical uncertainties of the cosmological models themselves will not allow unique tests to be performed to check the validity of one of the models. There will always be alternative ways to interpret the observational results in cosmology.

If, for example, we find a peculiar property of galaxies at, say, a redshift $z = 3$, we would need to go further back in time to higher redshifts to see whether this property persists to earlier times. But going further back will entail greater observational uncertainties since light from more distant sources will be blended together. To get around this problem we will have to make further theoretical assumptions which will not be directly testable. And this provides the central crux in all observations of cosmological import. The experience has shown that as our observations get better, we push our knowledge further and further back to times where theory itself cannot provide unequivocal predictions to carry out further tests.

If the flatness, horizon, and isotropy problems are really not problems for theory to explain, then the obvious question is why measurement leads to logically disparate constructs describing the universe. We are left here with a situation in which we seemingly cannot, in principle, determine whether the universe is open or closed, whether the big-bang or the steady-state model is correct, whether the redshifts are cosmological or not, or whether the constants of nature are "constant" or actually vary over time as first suggested by Dirac.[7] Recently, H. Alfven and others[8] have suggested that the universe might not have started from a big bang and that gravity does not dominate the large-scale structure of the universe. The hypothesis here is that electromagnetic effects in the cosmic plasma determine the structure of the universe including the observed superclusters. The following is a list of possible profound complementarities which, we believe, may be disclosed in our future studies of the large scale structure of the universe:

- open big-bang model/closed big-bang model
- big-bang cosmology/steady-state cosmology
- cosmological redshifts/discordant redshifts
- constants of nature are invariable/"constants" are varying
- big-bang cosmology/plasma cosmology

What we confront here, in our view, is not, however, a practical limit to observation. We are confronting rather conditions for observation comparable to those which result in wave-particle dualism. Here, as in laboratory experiments in quantum physics, we confront a situation in which the observer and his observing apparatus must be taken into account. In what follows, we concentrate on a particular complementary pair of cosmological import, the open and closed models in the big-bang cosmology. This discussion will illustrate the basic arguments which can be generalized to apply to the other candidates for complementary constructs in cosmology listed above.

Any big-bang model of the universe, open or closed, results in the observational horizon which applies to all events prior to a few hundred thousand years, or during the period when the universe was opaque to its own radiation. Any particular model adopted is ambiguous for earlier times, and an appeal to observations will not resolve the inherent ambiguities. It also seems clear that the adoption of a single and specific theoretical model requires the existence of this observational horizon which, nevertheless, *is not necessarily inherent or of the same type in different theoretical cosmological models.* If one, for example, adopts the steady state cosmological theory or Alfven's plasma cosmological theory, which assumes no beginning, and, therefore, no relic background radiation, one does not confront the big-bang problem of optical thickness at z = 1000. Similarly, we emphasize that if the redshifts are not cosmological, as Halton Arp, Geoffrey Burbidge, Fred Hoyle, and others have argued, this non-cosmological assumption would eliminate the z = 1000 optical thickness observation problem encountered in the big-bang models. Halton Arp[9] has found numerous examples where an apparent association between a galaxy of a small redshift and a quasar of a different and generally large (or "discordant") redshift occurs. These associations may be chance associations of nearby galaxies and more distant quasars that happen to be along the same lines of sight, even though a number of gaseous bridges have been found which would indicate spatial associations. Physicist Emil Wolf[10] has proposed a plausible alternate physical mechanism to explain galaxy redshifts. Even though these alternate ideas to the big-bang are not in vogue in view of the established big-bang paradigm, one has to remember that the big-bang theory is just a theory and not a self-evident truth. It is, therefore, subject to observational scrutiny. Yet these other models also present their own horizons of observability which are not the same as those encountered in the big-bang models.

Returning to the open/closed complementary pair, we are suggesting that, in an analogous situation to what one encounters in the microscopic world, these two models should be considered as complementary aspects of the complete big-bang theory of the universe. It is already clear in cosmology that the observations yield results which can be used to support each of these logically antithetical models. When one attempts to add up all luminous matter in the universe, one obtains insufficient mass to "close" the universe—the universe appears, therefore, to be open. When, on the other hand, one attempts to study the behavior of the Hubble Law for distant galaxies, one finds a tendency towards a closed model. Although the errors in this second attempt are large, we are predicting that future improved observations will not resolve the existing ambiguity. In fact, what distinguishes our approach from other theoretical approaches which attempt to explain the observations by forcing theoretical models, is that we are predicting that the ambiguities will not go away. They will remain as long as one insists on attempting to explain _all_ observations within a single model.

Quantum Resolution of the Cosmological Measurement Problems

If we can divest ourselves of the distorting lenses of classical assumptions, the case that the universe is a quantum system at all scales and times is easily made. If this conclusion is as factual as we understand it to be, we can then proceed to address the problems encountered by the big-bang cosmology in terms of first principles, as opposed to appealing to an ad hoc model like inflation. There is, first of all, general agreement among virtually all cosmologists and quantum physicists alike that all quanta were entangled early on in the history of the universe. If that is the case, we must also conclude that this _entanglement should remain a frozen-in property of the macrocosm._[11] Since quantum entanglement in the experiments testing Bell's theorem reveals an underlying wholeness that remains a property of the entire system even at macroscopic distances, the seemingly inescapable conclusion is that the underlying wholeness associated with quantum entanglement in the early universe remains a property of the universe at all times and all scales. If this is the case, it would seem to imply that any subsequent interaction with quanta at any stage in the life of the universe could reveal emergent complementarities.

In attempting to prove the "correctness" of one of the disparate constructs, we have to observe increasingly distant sources of light. Bigger and bigger ground telescopes and space telescopes are being built to collect the few photons expected from distant, faint sources. One cannot, however, ultimately avoid the problems arising from the complementary nature of light: If the expected photons are few, one would have to make the following choice: 1) disperse the light to obtain a spectrum, which would then yield the cosmological distance in which case the exact location of the faint source in the sky would be indeterminate; or 2) photographically record the position of the faint source in which case its spectrum, and distance, would be unknown. Someone may object here that two observations can be carried out, one in which the spectrum is obtained followed by a second observation in which the position is obtained. Yet, due to the faintness of the presumed source of light and the apparent nearness to it of other faint sources of light, one could not be sure that the photons in the second experiment emanated from the same exact location in the sky as the photons in the first experiment. Dispersing more of the light to obtain a better spectrum would exacerbate the problem, as the uncertainty principle implies, because the direction of origin of the observed photons in the sky would then become more uncertain. If the quantum nature of light has to be taken into account in observations of cosmological import and if the entanglement of quanta in the early universe remains a frozen-in property of a fundamentally quantum universe, one would not be surprised to encounter horizons of knowledge. Reaching these horizons would entail the requirement to apply the principle of complementarity. Assuming this possibility, we again emphasize that it is not unreasonable to conclude that the flatness, horizon, and isotropy problems of the big-bang models could be the direct result of limits of observation in a quantum universe which call for the application of *complementary* theoretical constructs. Perhaps what makes the observations problematic is that cosmologists have treated them as preconditions for *the confirmation of single* theoretical models which compete with one another for complete domination of the situation. These "problems," and others which future observations in cosmology are likely to reveal, could, therefore, be eliminated with the simple realization that quantum entanglement remains a frozen-in property of the cosmos, even at "macroscopic" scales, and for all times.

One can then conclude that these problems arise as a result of making observational choices in observing a quantum system, and are not really problems. Perhaps what we are actually confronting here is merely additional demonstrations of the underlying wholeness implicit in this system. In our view, *the reason the universe is so close to the flat, Euclidean model is that this model represents the division between the complementary constructs of open and closed cosmologies.* As our observations carry us to fainter and fainter sources, which in the big-bang model correlates with more distant regions of space, the observational uncertainties are merely testifying to our inability to apply a single model where another, logically disparate but complementary model, is also required for a complete view of the situation. However one feels about this argument, and some feelings are likely to be strongly negative, the observational evidence in cosmology suggests that there will "always" be alternative ways to interpret the results. Experience has shown that as our observations get better, we push our knowledge further and further back to times where theory itself cannot provide unequivocal predictions to carry out further tests.

One also faces the problem of initial conditions for the universe. Attempts by Stephen Hawking and others to circumvent this problem by appealing to the "no boundary" proposal ontologize the wave function of the universe and are inadmissible from the perspective of epistemological realism as we define it. Finally, what seems reasonable to conclude at the moment is that observational horizons of knowledge have left us in a situation in which we seemingly cannot, in principle, determine whether the universe is open or closed. The majority view is probably that we have merely confronted a practical limit to observation which we can easily imagine how to resolve by future observations. Our alternative view is that what we are actually confronting are conditions for observation comparable to those which invoke wave-particle dualism in quantum physics. We emphasize that the reason that we cannot decide whether the open or closed model is correct becomes simple and obvious—the logically disparate constructs are complementary. Although one precludes the other in application to a particular situation, both are needed for the complete description. In other words, that which is prior to any particular measurement of this reality is the same thing that is prior to measurement in all quantum mechanical experiments—the undivided wholeness of reality-in-itself. Yet the wholeness in the universe we have in mind here is not an a priori philosophical assumption—it is rather an emergent property of a quantum universe which reveals itself under experimental conditions that clearly indicate that the observer cannot be independent of the observing process *at any level.*

Since we needed a more sophisticated Theory to explain wave-particle duality —

so also we need a more sophisticated Theory of Cosmology to explain both open or close model of the universe

model theory ─── condition #1 ─── ▷ open
                 condition #2 ─── ▷ closed

Notes

Introduction

1. See A. Aspect, P. Grangier, and G. Roger, *Physical Review Letters*, 1981, *47*, p. 460; A. Aspect, P. Grangier, and G. Roger, *Physical Review Letters*, 1982, *49*, p. 91; A. Aspect, J. Dalibard, and G. Roger, *Physical Review Letters*, 1982, *49*, p. 1804.
2. John. S. Bell, *Physics*, 1964, *1*, p. 195.
3. Proceedings of Conference held at George Mason University, October 21–22, 1988, *Bell's Theorem, Quantum Theory and Conceptions of the Universe*, ed. Menas Kafatos (Heidelberg: Kluwer Academic Press, 1989).
4. Frederick Suppe, *The Structure of Scientific Theories* (Chicago: The University of Illinois Press, 1977).
5. Ibid, p. 126.
6. See S. Toulmin, *Foresight and Understanding* (New York: Harper and Row Torchbook, 1963).
7. Frederick Suppe, *The Structure of Scientific Theories*, p. 128.
8. S. Toulmin, *Foresight and Understanding*, pp. 53, 77.
9. Frederick Suppe, *The Structure of Scientific Theories*, p. 131.
10. Ibid, p. 133.
11. Thomas S. Kuhn, *The Structure of Scientific Revolutions* (Chicago: The University of Chicago Press, 1970), p. 91.
12. Ibid, p. 10.
13. See Thomas S. Kuhn, "Reflections on My Critics," in *Criticism and Growth of Knowledge*, eds. O. Lakatos and A.E. Musgrave (Cambridge: Cambridge University Press, 1970), pp. 231–278.
14. Ibid, p. 232.
15. Frederick Suppe, *The Structure of Scientific Theories*, p. 165.
16. N.R. Hanson, *Patterns of Discovery* (Cambridge: Cambridge University Press, 1958), p. 71.
17. Frederick Suppe, *The Structure of Scientific Theories*, p. 152.
18. P.K. Feyerabend, *Against Method: Outlines of an Anarchistic Theory of Knowledge* (London: New Left Books, 1975), p. 21.
19. Frederick Suppe, *The Structure of Scientific Theories*, pp. 650–651.
20. Ibid, pp. 694–695.
21. Stephen W. Hawking, *A Brief History of Time* (New York: Bantam, 1988), pp. 140–141.
22. Abner Shimony, "Our World and Microphysics," in *Philosophical Consequences of Quantum Theory*, eds. James T. Cushing and Ernan McMullin, (Notre Dame: University of Notre Dame Press, 1989), pp. 37–25.

Chapter 1

1. See Ivor Leclerc, "The Relation Between Science and Metaphysics," in *The World View of Contemporary Physics*, ed. Richard F. Kitchener (Albany: S.U.N.Y Press, 1988), p. 28.

2. Albert Einstein, "Aether and Relativitatstheorie," trans. W. Perret and G.B. Jeffrey, in *Physical Thought from the Pre-Socratics to the Quantum Physicists*, ed. S. Samsbursky (New York: Pica Press, 1975), p. 479.
3. Ernest Rutherford, quoted in Ruth Moore, *Niels Bohr: The Man, His Science and the World They Changed* (New York: Knopf, 1966), p. 21.
4. Werner Heisenberg, quoted in James B. Conant, *Modern Science and Modern Man* (New York: Columbia University Press, 1953), p. 40.
5. Max Jammer, *The Conceptual Development of Quantum Mechanics* (New York: McGraw Hill, 1966), p. 271.
6. Werner Heisenberg, quoted in James B. Conant, *Modern Science and Modern Man*, p. 271.
7. Robert Oppenheimer, quoted in Ibid.
8. Clifford A. Hooker, "The Nature of Quantum Mechanical Reality," in *Paradigms and Paradoxes*, ed. Robert G. Colodny (Pittsburgh: University of Pittsburgh Press, 1972), p. 132.

Chapter 2

1. Eugene P. Wigner, "The Problem of Measurement," in *Quantum Theory and Measurement*, eds. John A. Wheeler and Wojciech H. Zurek (Princeton: Princeton University Press, 1983), p. 327.
2. Ibid, p. 327.
3. Olivier C. de Beauregard, private communication.
4. Richard Feynman, *The Character of Physical Law* (Cambridge, Massachusetts: MIT Press, 1967), p. 130.
5. John A. Wheeler, "Beyond the Black Hole," in *Some Strangeness in the Proportion*, ed. Harry Woolf (London: Addison-Wesley, 1980), p. 354.
6. See Abner Shimony, "The Reality of the Quantum World," *Scientific American*, January 1988, p. 46.
7. Richard P. Feynman, *QED: The Strange Theory of Light and Matter* (Princeton: Princeton University Press, 1985), p. 7.
8. Ibid, p. 25.
9. See, for example, Paul C.W. Davies, *Quantum Mechanics* (London: Routledge & Kegan Paul, 1984).
10. Richard P. Feynman, *The Character of Physical Law*, p. 80ff.
11. Abner Shimony, "The Reality of the Quantum World," p.48.
12. Ibid.
13. Steven Weinberg, quoted in Heinz Pagels, *The Cosmic Code* (New York: Bantam Books, 1983), p. 239.
14. Errol E. Harris, "The Universe in the Light of Contemporary Scientific Developments," in *Bell's Theorem, Quantum Theory and Conceptions of the Universe*, pp. 305–311.

Chapter 3

1. Abraham Pais, *Subtle Is the Lord* (New York: Oxford University Press, 1982).

2. A. Einstein, B. Podolsky, and N. Rosen, "Can Quantum-Mechanical Description of Physical Reality Be Considered Complete?," *Physical Review,* 1935, *47*, p. 777. Paper is reprinted in *Physical Reality*, ed. S. Toulmin (New York: Harper and Row, 1970).
3. Albert Einstein, Ibid.
4. Quoted in Abraham Pais, *Subtle Is the Lord*, p. 456.
5. Nick Herbert, *Quantum Reality: Beyond the New Physics, An Excursion into Metaphysics and the Meaning of Reality*, p. 216ff.
6. See A. Aspect, J. Dalibard, and G. Roger, *Physical Review Letters,* 1981, *47*, p. 460.
7. Bernard d'Espagnat, *Physical Review Letters,* 1981, *49*, p. 1804.
8. See Gary Zukav, *The Dancing Wu Li Masters* (New York: Bantam New Age, 1979), p. 299.
9. See Bernard d'Espagnat, *In Search of Reality* (New York: Springer-Verlag, 1981), pp. 43–48.

Chapter 4

1. Ivor Leclerc, "The Relation Between Science and Metaphysics," in *The World View of Contemporary Physics*, ed. Richard E. Kitchener (Albany: S.U.N.Y. Press, 1988), p. 30.
2. Ibid, p. 27.
3. Ibid, p. 28.
4. Ibid.
5. Ibid.
6. Ibid, p. 29.
7. Ibid, p. 31.
8. Gerald Holton, "Do Scientists Need Philosophy?" *The Times Literary Supplement*, November 2, 1984, pp. 1231–1234.
9. Ivor Leclerc, "The Relation Between Science and Metaphysics," pp. 25–37.
10. Albert Einstein, *The World as I See It* (London: John Lane, 1935), p. 134.
11. Ibid, p. 136.
12. Ivor Leclerc, "The Relation Between Science and Metaphysics," p. 31.
13. Niels Bohr, *Atomic Theory and the Description of Nature* (Cambridge: Cambridge University Press, 1961), pp. 4, 34.
14. See Clifford A. Hooker, "The Nature of Quantum Mechanical Reality," in *Paradigms and Paradoxes*, pp. 161–162. Also see Niels Bohr, *Atomic Physics and Human Knowledge*, (New York: John Wiley and Sons, 1958), pp. 34, 26, 72, and 88ff, and *Atomic Theory and the Description of Nature*, pp. 5, 8, 16ff, 53, and 94.
15. Herman Feshbach and Victor F. Weisskopf, "Ask a Foolish Question," *Physics Today*, October 1988, pp. 9–11.
16. Niels Bohr, "Causality and Complementarity," *Philosophy of Science, 4*, pp. 293–294.
17. Ibid.
18. Clifford A. Hooker, "The Nature of Quantum Mechanical Reality," p. 137.
19. See Abraham Pais, *Subtle Is the Lord*.
20. Leon Rosenfeld, "Niels Bohr's Contributions to Epistemology," *Physics Today*, April 29, 1961, *190*, p. 50.
21. Niels Bohr, *Atomic Theory and the Description of Nature*, pp. 54–55.
22. Niels Bohr, "Discussions with Einstein on Epistemological Issues," in Henry Folse *The Philosophy of Niels Bohr: The Framework of Complementarity* (Amsterdam: North Holland Physics Publishing, 1985), pp. 237–238.

23. Niels Bohr, *Atomic Physics and Human Knowledge*, pp. 64, 73. Also see Clifford Hooker's detailed and excellent discussion of these points in "The Nature of Quantum Mechanical Reality," in *Paradigms and Paradoxes* (Pittsburgh: University of Pittsburgh Press, 1972), pp. 57–302.
24. Clifford A. Hooker, "The Nature of Quantum Mechanical Reality," p. 155.
25. Niels Bohr, *Atomic Physics and Human Knowledge*, p. 74.
26. Niels Bohr, *Atomic Theory and the Description of Nature*, pp. 56–57.
27. Niels Bohr, *Atomic Physics and Human Knowledge*, p. 74.
28. Niels Bohr, "Physical Science and Man's Position," *Philosophy Today,* 1957, p. 67.
29. Leon Rosenfeld, "Foundations of Quantum Theory and Complementarity," *Nature,* April 29, 1961, *190,* p. 385.
30. Niels Bohr, *Atomic Physics and Human Knowledge*, p. 79.
31. Niels Bohr, quoted in A. Peterson, "The Philosophy of Niels Bohr," *Bulletin of the Atomic Scientists,* September 1963, p. 12.
32. Niels Bohr, *Atomic Theory and the Description of Nature*, p. 49.
33. Albert Einstein, *Ideas and Opinions* (New York: Dell, 1976), p. 271.
34. See Melic Capek, "Do the New Concepts of Space and Time Require a New Metaphysics," in *The World View of Contemporary Physics*, pp. 90–104.
35. Fritjof Capra, "The Role of Physics in the Current Change of Paradigms," in *The World View of Contemporary Physics*, p. 153.
36. Niels Bohr, quoted in A. Peterson, "The Philosophy of Niels Bohr," p. 12.
37. Henry P. Stapp, "S Matrix Interpretation of Quantum Theory," *Physical Review,* 1971, *3,* 1303 ff.
38. Henry J. Folse, "Complementarity and Space-Time Descriptions," in *Bell's Theorem, Quantum Theory and Conceptions of the Universe*, p. 258.

Chapter 5

1. Karl Mannheim, *Ideology and Utopia: An Introduction to the Sociology of Knowledge* (London: Routledge and Kegan Paul, 1936). pp. 261–268.
2. See Ludwig Wittgenstein, *Philosophical Investigations*, trans. G. E. M. Anscombe (Oxford: Blackwell, 1953).
3. See P.L. Berger and T. Luckmann, *The Social Construction of Reality: A Treatise in the Sociology of Knowledge* (New York: Penguin, 1967).
4. David Bloor, *Knowledge and Social Imagery* (London: Routledge and Kegan Paul, 1967); and Bruno Latour and Steve Woolgar, *Laboratory Life: The Social Construction of Scientific Facts* (Beverly Hills: Sage, 1979).
5. Stephen Jay Gould, *Ever Since Darwin* (New York: Norton, 1977), p. 12.
6. Ernest Mayer, quoted in Rueben Abel, *Man Is the Measure* (New York: Macmillan, 1976), p. 99.
7. Richard Gregory, *The Intelligent Eye* (New York: McGraw Hill, 1970), p. 45.
8. See Sue Taylor Parker and Kathleen Rita Gibson, "A Developmental Model for the Evolution of Language and Culture in Early Hominids," *Behavioral and Brain Sciences,* 1979, *2*(3).
9. Philip Tobias, quoted in Seymour W. Itzkoff, *The Form of Man* (Ashfield, Massachusetts, 1983), p. 212.

10. See Alan Walker and Richard E.F. Leaky, "The Hominids in East Turkana," *Scientific American*, August 1978, pp. 54–56; Donald Johanson and T. D. White, "A Systematic Assessment of Early African Hominids," *Science*, 1970, *203*, pp. 321–330; and John E. Cronin et al., "Tempo and Mode of Hominid Evolution," *Nature*, March 1981, *291*, pp. 113–122.

11. See Richard Gregory, *Concepts and Mechanisms of Perception* (London: Gerald Duckworth, 1974); *The Intelligent Eye* (London: Wiedenfeld and Nicholson, 1970).

12. Clifford Geertz, quoted in *Man Is the Measure*, p. 154.

13. Ludwig Wittgenstein, quoted in *Man is the Measure*, p. 155.

14. Walter F. Otto, *The Homeric Gods*, trans. Moses Hadas (New York: Vintage Books, 1954), pp. 6–7.

15. Copernicus, *De Revolutionibus*, quoted in Gerald Holton, *Thematic Origins of Modern Thought* (Cambridge, Massachusetts: Harvard University Press, 1974), p. 82.

16. Kepler to Hewart von Hohenberg, quoted in Ibid, p. 76.

17. Alexander Koyre, *Metaphysics and Measurement* (Cambridge, Massachusetts: Harvard University Press, 1968), pp. 42–43.

18. Heinrich Hertz, quoted in Heinz Pagels, *The Cosmic Code* (New York: Basic Books, 1983), p. 301.

19. Albert Einstein, "Autobiographical Notes," in *Albert Einstein: Philosopher-Scientist*, ed. P.A. Schlipp (New York: Harper and Row, 1959), p. 210.

20. Albert Einstein, "On the Method of Theoretical Physics," in *Ideas and Opinions* (New York: Dell, 1973), pp. 246-247.

21. Gerald Holton, *Thematic Origins of Modern Science*, p. 307.

22. Ilse Rosenthal-Schneider, "Reminiscences of Conversations with Einstein," July 23, 1959, quoted in Ibid., p. 236.

23. Albert Einstein, in Abraham Pais, *Subtle Is the Lord* (New York: Oxford University Press, 1982), p. 115.

Chapter 6

1. Steven Weinberg, quoted in Paul Davies, *The Superforce* (New York: Simon and Schuster, 1984), p. 222.

2. Jacques Monod, quoted in Ilya Prigogine and Isabelle Stengers, *Order Out of Chaos* (New York: Bantam Books, 1984), p. 187.

3. Gerald Feinberg, quoted in Heinz Pagels, *The Cosmic Code* (New York: Bantam Books, 1983), p. 187.

4. Erwin Schrödinger, quoted in *Order Out of Chaos*, p. 18.

5. Evelyn Fox Keller, "Cognitive Repression in Contemporary Physics, *American Journal of Physics*, August 1979, *47*(8), p. 717.

6. Henry P. Stapp, "Quantum Theory and the Physicist's Conception of Nature: Philosophical Implications of Bell's Theorem, in *The World View of Contemporary Physics*.

7. Max Planck, *Where Is Science Going?* (London: G. Allen and Unwins, 1933), p. 24.

8. Albert Einstein, "Autobiographical Notes," in *Albert Einstein: Philosopher-Scientist*, ed. P.A. Schlipp (New York: Harper and Row, 1959), p. 3.

9. Albert Einstein, quoted in *New York Post*, November 28, 1972.

10. Henry P. Stapp, "Quantum Theory and the Physicist's Conception of Nature: Philosophical Implications of Bell's Theorem," in *The World View of Contemporary Physics*, p. 40.

11. See Henry P. Stapp, "Quantum Ontologies," in *Bell's Theorem, Quantum Theory and Conceptions of the Universe*, pp. 269–278.
12. Ibid.
13. Ibid.
14. See David Bohm, *Wholeness and the Implicate Order* (London: Routledge and Kegan Paul, 1980).
15. Henry P. Stapp, "Quantum Ontologies," p. 273.
16. Ibid.
17. Ibid, p. 275
18. Ilya Prigogine and Isabelle Stengers, *Order Out of Chaos* (New York: Bantam Books, 1984), p. 76.
19. Alexander Koyre, *Metaphysics and Measurement* (Cambridge, Massachusetts: Harvard University Press, 1968), pp. 42–43.
20. Alexander Koyre, *Newtonian Studies* (Chicago: University of Chicago Press, 1968), pp. 23–24.
21. See account and comments on sources in G. Milhaud, *Descartes Savant* (Paris, 1922).
22. Rene Descartes, *Principles of Philosophy*, Part I, principle 26, in *Philosophical Works*, trans. Haldene and Ross (Cambridge, Massachusetts: Harvard University Press, 1911).
23. Ibid.
24. Martin Heidegger, quoted in George Steiner, *Martin Heidegger* (New York: Viking Press, 1978), p. 25.
25. Martin Heidegger, quoted in Ibid.
26. David Bohm, "Interview," *Omni*, January 1987, *9*(4), pp. 69–74.

Chapter 7

1. Leon Rosenfeld, "Niels Bohr in the Thirties: Consolidation and Extension of Complementarity," in *Niels Bohr: His Life and Work As Seen By His Friends and Colleagues*," ed. Stephan Rozental (New York: John Wiley and Sons, 1964), p. 121.
2. Niels Bohr, *Atomic Physics and Human Knowledge* (New York: John Wiley and Sons, 1958), p. 91.
3. Niels Bohr, *Atomic Theory and the Description of Nature* (Cambridge: Cambridge University Press, 1961), p. 100.
4. Ibid.
5. Ibid.
6. Henry J. Folse, *The Philosophy of Niels Bohr*, (Amsterdam: North Holland Publishing, 1985) p. 174ff.
7. Niels Bohr, "Interview," in *Archive for the History of Quantum Physics*, Thomas Kuhn et al., November 17, 1962, p. 2. [Also see Henry J. Folse, *The Philosophy of Niels Bohr*, pp. 176–177.]
8. Niels Bohr, "Interview," in Ibid.
9. See Noam Chomsky, *Cartesian Linguistics* (New York: Harper and Row, 1966); and *Language and Mind* (New York: Harcourt, 1968).
10. Ferdinand de Saussure, *Course in General Linguistics*, trans. Roy Harris (London: Duckworth, 1983).
11. Ibid, p. 110.
12. Michael Foucault, *The Order of Things: An Archeology of the Human Sciences* (New York: Vintage-Random House, 1973).
13. Jacques Lacan, *Ecrits: A Selection*, trans. A. Sheridan (New York: Norton, 1977), p. 145.

14. Ibid, p. 150.
15. Jacques Darrida, *Margins of Philosophy*, trans. A. Bass (Chicago: University of Chicago Press, 1982), p. 11.
16. Ibid, p. 23.
17. Ibid, pp. 3–4.
18. Ibid, p. 22.
19. A.J. Greimas, *Semantiques Structurale* (Paris: Larousse, 1966).
20. Claude Levi-Strauss, *Le Cru et le cuit*, in *Structuralism*, ed. Jacques Ehrmann (Garden City, New York: Anchor-Doubleday, 1970).
21. Henry P. Stapp, "Quantum Theory of Consciousness," unpublished paper, p. 5–6.
22. See Richard Restak, *The Brain* (New York: Warner Books, 1970), pp. 52–55.
23. See Niels Bohr, *Atomic Physics and Human Knowledge*.
24. Niels Bohr, "Biology and Atomic Physics," in Ibid, pp. 20–21.
25. Rupert Sheldrake, *A New Science of Life* (Los Angeles: J.P. Tarcher, 1981).
26. Harold Morowitz, private communication.
27. Ilya Prigogine and Isabelle Stengers, *Order Out of Chaos* (New York: Bantam Books, 1984).
28. Ibid, pp. 12–14, 142–143.
29. It is interesting to note that the "horizon" which separates these two complementary constructs has to do with parameters taking on different mathematical values.

Chapter 8

1. David N. Schramm, "The Early Universe and High-Energy Physics," *Physics Today*, April 1983, p. 27.
2. R. Brent Tully, "More about Clustering on a Scale of 0.1 c," *Astrophysical Journal*, 1987, *323*, p. 1–18.
3. See report on "Physics News," *Physics Today*, January 1988.
4. Alan H. Guth and Paul J. Steinhardt, "The Inflationary Universe," *Scientific American*, May 1984, p. 116. Also, John D. Barrow, *Q. Jl R. Astr. Soc.*, 1988, *29*, pp. 101–117.
5. Ibid.
6. See Menas Kafatos and Robert Nadeau, "Complementarity and Cosmology," p. 261, and Menas Kafatos, "Horizons of Knowledge in Cosmology," in *Bell's Theorem, Quantum Theory and Conceptions of the Universe*, p. 195.
7. Ibid.
8. John A. Wheeler, "Law without Law," in *Quantum Theory and Measurement*, p. 182.
9. Menas Kafatos and Thalia Kafatou, *Undivided Wholeness: The Quantum, the Universe and Consciousness* (book manuscript).
10. See, for example, Menas Kafatos, "The Universal Diagrams and Life in the Universe," in *The Search for Extraterrestrial Life: Recent Developments*, ed. Michael D. Papagiannis (Dordrecht: D. Reidel Co., 1985), pp. 245–249.
11. Ibid. Also see Menas Kafatos, "The Position of Brown Dwarfs on the Universal Diagrams," in *Astrophysics of Brown Dwarfs*, ed. M. Kafatos (Cambridge: Cambridge University Press, 1986), p. 198.

12. The term "bifurcation" is known from chaos theory as a transition is made from well-behaved ordered states to chaotic states. As one approaches the point of bifurcation, small changes in a relevant mathematical parameter can throw the system into an ordered or chaotic system. This transition is also evidenced in the fractal geometry where small changes in the dimensionality lead to totally different patterns. Such a bifurcation process may indeed be the driving force for the emergence of complementarities at all scales and levels and for the emergence of patterning structures in the universe. An international conference titled "Patterns in the Universe" which examined what constitutes a pattern in various scientific, mathematical and philosophical fields was organized at the Smithsonian Institution by one of the authors (M.K.). Bifurcation processes were evident in many topics. One should compare these ideas to A. N. Whitehwead's "bifurcation of nature" (see Abner Shimony, "Our Worldview and Microphysics," p. 37).

13. Roger Penrose, *The Emperor's New Mind* (New York: Oxford University Press, 1989).

Chapter 9

1. Werner Heisenberg, *Quantum Questions,* ed. Ken Wilbur (Boulder: New Science Library, 1984), p. 44.
2. Henry P. Stapp, "Quantum Theory and the Physicist's Conception of Nature: Philosophical Implications of Bell's Theorem, " in *The World View of Contemporary Physics*, p. 38.
3. Ibid.
4. Melic Capek, "New Concepts of Space and Time," in Ibid, p. 99.
5. Henry P. Stapp, "Quantum Theory and the Physicist's Conception of Nature: Philosophical Implications of Bell's Theorem," in Ibid, p. 54.
6. Errol E. Harris, "Contemporary Physics and Dialectical Holism," in Ibid, p. 159.
7. Werner Heisenberg, *Physics and Philosophy* (London: Faber, 1959), p. 96.
8. Rudy Rucker, *Infinity and the Mind*, p. 157 (Boston: Birkhauser 1982).
9. Errol E. Harris, "Contemporary Physics and Dialectical Holism," in *The World View of Contemporary Physics*, p. 161.
10. Ibid.
11. Ibid.
12. Ibid, p. 162.
13. Ibid, p. 163.
14. Ibid, p. 164.
15. Ibid.
16. Ibid, p. 165.
17. Ibid.
18. Ibid, p. 171.
19. David Bohm, *Wholeness and the Implicate Order* (London: Routeledge and Kegan Paul, 1980), p. 210.
20. Werner Heisenberg, in *Quantum Questions*, p. 96.
21. Karl Jaspers, *Tragedy Is Not Enough*, trans. Reiche, Moore, and Deutsch (Boston: Anchor Books, 1952), p. 48.
22. Albert Einstein, in *Quantum Questions*, p. 111.
23. Jonas Salk, *Survival of the Wisest* (New York: Harper and Row, 1973), p. 82.
24. Fritjof Capra, "The Role of Physics in the Current Change of Paradigms," in *The World View of Contemporary Physics*, p. 151.

25. Ludwig von Bertalanffy, *General Systems Theory: Foundations, Development, Applications* (New York: Braziller, 1968), p. 13.

26. Anthony Wilden, *System and Structure: Essays in Communication and Exchange*, 2nd ed. (New York: Tavistock, 1980), p. 241.

27. Ibid. See also Ervin Laszlo, "The Emergence of Complexity: The Search for Enduring Patterns in the Universe," unpublished paper presented at the Conference "Patterns in the Universe", and Ervin Laszlo, *Cosmic Connections: Steps to the Theory of the Whole* (New York, Bantam Books, 1990). He is among the systems philosophers who have pointed out the importance of information and negative entropy in any theorical attempt to understand the universe. He points out that theories "predicting the irreversible loss of order and complexity may be incomplete," and suggests that the conservation of form or pattern in evolution implies the presence of memory on which all buildup of order and complexity is based. This "memory" should not, he emphasizes, be regarded in anthropomorphic terms, and parallels our thesis of a conscious universe which discloses complementary relationships.

28. Susan Stewart, *On Longing: Narratives of the Miniature, the Gigantic, the Souvenir, the Collection* (Baltimore: Johns Hopkins University Press, 1984).

29. Michael Serres, *The Parasite*, trans. Lawrence R. Schehr (Baltimore: Johns Hopkins University Press, 1982), pp. 72–73.

30. Wolfgang Pauli, in *Quantum Questions*, p. 163.

31. Erwin Schrödinger, in *Quantum Questions*, p. 97.

32. See also Menas Kafatos and Thalia Kafatou, *Undivided Wholeness: The Quantum, the Universe and Consciousness*.

33. Erwin Schrödinger, in *Quantum Questions*, p. 81.

Appendix

1. Menas Kafatos, "Horizons of Knowledge in Cosmology," in *Bell's Theorem, Quantum Theory and Conceptions of the Universe*, pp. 195–210.

2. Ibid.

3. See David N. Schramm, "The Early Universe and High-Energy Physics."

4. George Miley, *Physics World*, 1989, *2*, p. 35.

5. K.C. Chambers et al., *Nature*, 1987, *329*, p. 604. K.C. Chambers et al., *Astrophysical Journal (Letters)*, 1988, *329*, p. L75.

6. Menas Kafatos, "Horizons of Knowledge in Cosmology," pp. 195-210.

7. See, for example, discussion in Edward R. Harrison, "The Cosmic Numbers," *Physics Today*, December 1972, p. 30.

8. Anthony L. Peratt, *The Sciences*, January/February 1990, p. 24.

9. Halton Arp, *Astrophysical Journal (Letters)*, 1984, *277*, p. L27.

10. Emil Wolf, see, for example, *Science News*, 1989, *136*, p. 326.

11. Menas Kafatos, "Horizons of Knowledge in Cosmology," p. 209.

Index